REA: THE TEST PREP AP® TEACHERS RECOMMEN

AP® CHEMISTRY
CRASH COURSE®

Adrian Dingle

The Westminster Schools
Atlanta, Georgia

Research & Education Association
Visit our website at: www.rea.com

Research & Education Association
61 Ethel Road West
Piscataway, New Jersey 08854
E-mail: info@rea.com

AP® CHEMISTRY CRASH COURSE®

Published 2016

Printed in the United States of America

Library of Congress Control Number 2013954713

ISBN-13: 978-0-7386-1154-9
ISBN-10: 0-7386-1154-9

Cover image: Todd Warnock/Photodisc/Getty Images

AP CHEMISTRY CRASH COURSE TABLE OF CONTENTS

ABOUT THIS BOOK

REA's ***AP Chemistry Crash Course*** is the first book of its kind for the last-minute studier or any AP student who wants a quick refresher on the course. Our review and online practice test are based on a careful analysis of the AP Chemistry Course Description outline and actual AP test questions.

Written by an AP teacher, our easy-to-read format gives students a crash course in Chemistry. The targeted review chapters are grouped by topics, offering you a concise way to learn all the important facts, formulas, and terms before exam day.

Unlike other test preps, REA's ***AP Chemistry Crash Course*** gives you a review specifically focused on what you really need to study in order to ace the exam. The introduction discusses the content and format of the new exam, shows you the types of questions you'll see on test day, and explains how the exam is scored.

Parts Two through Seven give you a complete crash course in Chemistry. Each chapter presents essential information tested on the AP exam and covers everything from Atoms and Elements to Chemical Thermodynamics and Equilibrium. The author includes a chapter detailing the laboratory work necessary for AP Chemistry, and also shows you how to write high-scoring answers for the free-response portion of the exam.

No matter how or when you prepare for the AP Chemistry exam, REA's ***Crash Course*** will show you how to study efficiently and strategically, so you can boost your score!

To check your test readiness for the AP Chemistry exam, either before or after studying this ***Crash Course***, take REA's **FREE online practice exam**. To access your practice exam, visit the online REA Study Center at *www.rea.com/studycenter* and follow the on-screen instructions. This true-to-format test features automatic scoring of the multiple-choice questions, detailed explanations of all answers, and diagnostic score reporting that will help you identify your strengths and weaknesses so you'll be ready on exam day!

Good luck on your AP Chemistry exam!

ABOUT OUR AUTHOR

Adrian Dingle is a chemistry educator and author, with 24 years of experience teaching in the United States and the United Kingdom. He is the creator of the award-winning chemistry website, *www.adriandingleschemistrypages.com.*

The focus of Mr. Dingle's teaching career has been on preparing students for standardized tests; AP and SAT subject tests in the United States, GCSE's and A levels in the United Kingdom, and International Baccalaureate in both countries. An Englishman, he lives in Atlanta, Georgia, where he teaches at The Westminster Schools. He holds a B.Sc. (Hons.) Chemistry, and a Postgraduate Certificate in Education, both from the University of Exeter in England.

In addition to writing this *Crash Course*, Mr. Dingle has written *The Periodic Table: Elements With Style, How To Make A Universe With 92 Ingredients*, and REA's *SAT Chemistry Crash Course*. He is the 2011 winner of the School Library Association of the UK's Information Book Award, and, in 2012, was honored with the prestigious literary prize *Wissenschaftsbuch des Jahre,* sponsored by the Austrian Ministry of Science and Research.

"I would like to thank Paul Cohen, AP Chemistry teacher at Yeshivah of Flatbush High School, Brooklyn, New York, for his invaluable help and direction in preparing this book."—Adrian Dingle

ACKNOWLEDGMENTS

In addition to our author, we would like to thank Larry B. Kling, Vice President, Editorial, for his overall guidance, which brought this publication to completion; Pam Weston, Publisher, for setting the quality standards for production integrity and managing the publication to completion; Diane Goldschmidt, Managing Editor, for editorial project management; Alice Leonard, Senior Editor, for preflight editorial review; and Weymouth Design and Christine Saul, Senior Graphic Artist, for designing our cover.

We also extend our special thanks to Derrick C. Wood, Conestoga High School, Berwyn, Pennsylvania, for technically reviewing the manuscript; Ellen Gong for proofreading; and Kathy Caratozzolo of Caragraphics for typesetting this edition.

PART I

INTRODUCTION

Keys for Success on the AP Chemistry Exam

The textbooks that are typically used in college-level general chemistry courses contain thousands of facts and concepts packed into (at least) several hundred pages. If the AP Chemistry exam tested all of that information, earning a good score on the exam would be a daunting one.

Studying for the AP Chemistry exam requires you to be a pragmatic learner who can delineate the important (tested) material from the material that is "interesting to know." This book will help you to become more focused in your study, and help you to streamline your efforts. Your chances of scoring well on the exam will be enhanced by pinpointing what is absolutely necessary and ignoring the "fluff." The keys to success include the following:

1. The Content and Format of the AP Chemistry Exam

The Advanced Placement Chemistry curriculum is based on the content of an introductory chemistry course taught at the college level. The topics taught in the class reflect major topics that are presented in a number of college-level textbooks. The curriculum of the course also includes some inquiry-based laboratory situations, application of what the College Board calls *scientific practices* and chemical calculations.

In order to succeed on the exam, students need to master the basic concepts of chemistry and apply these concepts to various situations in a traditional test format. The content of the exam is based on the following six "big ideas":

1. The chemical elements are fundamental building materials of matter, and all matter can be understood in terms of

arrangements of atoms. These atoms retain their identity in chemical reactions. (See Part II: Atoms and Elements.)

2. Chemical and physical properties of materials can be explained by the structure and the arrangement of atoms, ions, or molecules and the forces between them. (See Part III: Bonding.)

3. Changes in matter involve the rearrangement and/or reorganization of atoms and/or the transfer of electrons. (See Part IV: Chemical Reactions.)

4. Rates of chemical reactions are determined by details of the molecular collisions. (See Part V: Rates of Reaction.)

5. The laws of thermodynamics describe the essential role of energy and explain and predict the direction of changes in matter. (See Part VI: Chemical Thermodynamics.)

6. Any bond or intermolecular attraction that can be formed can be broken. These two processes are in a dynamic competition, sensitive to initial conditions and external perturbations. (See Part VII: Equilibrium.)

The AP Chemistry exam is 3 hours in length and consists of a 90-minute multiple-choice question section and a 90-minute free-response question section. There are 60 multiple-choice questions and 7 free-response questions.

Section I. 60 multiple-choice questions (90 minutes) – 50% of the grade

In this section of the exam, NO calculator is allowed, but access to the periodic table and an equations and constants sheet is allowed.

Section II. 7 free-response questions (3 "long" questions and 4 "short" questions, with multiple parts) (90 minutes) – 50% of the grade

In this section of the exam, a calculator is allowed along with access to the periodic table and an equations and constants sheet.

2. **The Multiple-Choice Questions**

Multiple-choice questions test the knowledge, understanding, and application of the "big ideas" and science practices in the AP Chemistry course. For example:

When the chemical equation below is completed and balanced, what statement best describes the reaction?

$$C_5H_{12}(g) + O_2\,(g) \rightarrow ?$$

(A) 1 mole of pentane is consumed and 6 moles of water are produced.

(B) 1 mole of pentane is produced and 6 moles of water are consumed.

(C) 1 mole of pentane is consumed and 1 mole of water is produced.

(D) 1 mole of pentane is produced and 1 mole of water is consumed.

The correct answer is (A). The equation represents a combustion reaction that will produce carbon dioxide and water. When balanced the equation is:

$$C_5H_{12}(g) + 8O_2(g) \rightarrow 5CO_2(g) + 6H_2O(g)$$

It is essential to know these basic ideas when confronted by this type of question. This book will give you the content that you need to know to successfully answer the questions.

In addition, calculation-based chemistry questions are also part of this section. However, since no calculator is allowed, some multiple-choice questions will ask you to select the correct mathematical setup, or require you to do some simple arithmetic or estimation. For example:

At 25°C, 300 milliliters of an ideal gas exerts a pressure of 740 mm Hg. What is the volume of the gas under conditions of 0°C and 760 mm Hg?

(A) 2.68 mL

(B) 26.8 mL

(C) 268 mL

(D) 2680 mL

The correct answer is (C). Using the combined gas law formula, remembering to convert temperatures to Kelvin and plugging in values, the following mathematical setup is determined:

$$\frac{P_1V_1}{T_1} = \frac{P_2V_2}{T_2}, \quad \frac{(740)(300)}{298} = \frac{(760)(V_2)}{273}$$

Solving for V_2 gives $300 \times \frac{740}{760} \times \frac{273}{298}$ mL. Although you have no calculator, you should see that 300 mL is being multiplied by two numbers that are slightly less than 1 $\left(\frac{740}{760} \text{ and } \frac{273}{298}\right)$, and, as such, the answer must be slightly less than 300 mL; 268 mL is the correct answer choice. Estimation should allow you to find the correct answer without the need for a calculator (i.e., without the need to find the exact answer).

Multiple-Choice Question Formats

There are two types of multiple-choice questions that make up the majority of Section I of the AP Chemistry exam, with Type II making up a maximum of 50% of the exam.

Type I. Traditional multiple-choice questions with answer choices (A) through (D). There is only one correct answer. For example:

Atoms of element X have the electron configuration shown below.

$$1s^2 2s^2 2p^6 3s^2 3p^4$$

The compound most likely formed with lithium, Li, is

(A) LiX

(B) LiX_2

(C) Li_2X

(D) LiX_3

The correct answer is (C). A lithium ion has a charge of +1. Element X will have an ion with a charge of –2, and the ions must combine to give a compound with no overall charge.

Type II. Question sets. A question set, is a set of two to six questions that are preceded by a single stimulus or set of data, which usually includes a paragraph or two of text. Once again, there is only one correct answer for each question. For example:

Questions 26–28

The chemical reaction between sodium metal and chlorine gas can be represented by the equation shown below.

$$Na(s) + \frac{1}{2} Cl_2(g) \rightarrow NaCl(s)$$

ΔH = negative (i.e., it is an exothermic reaction)

The reaction can be broken down into a number of individual processes, including:

Atomization (sublimation) of sodium

First ionization energy of sodium

Atomization of chlorine

First electron affinity of chlorine

Lattice energy of sodium chloride

26. Which of the following represents the first electron affinity of chlorine?

(A) $Cl(g) + e^- \rightarrow Cl^-(g)$

(B) $Cl^-(g) + e^- \rightarrow Cl^{2-}(g)$

(C) $Cl(g) \rightarrow Cl^+(g) + e^-$

(D) $Cl_2(g) \rightarrow 2Cl(g)$

The correct answer is (A). The equation in (A) represents the definition of first electron affinity.

27. What is the expected energy change when chlorine gas is produced by the decomposition of NaCl?

(A) Exothermic, energy is released

(B) Exothermic, energy is absorbed

(C) Endothermic, energy is released

(D) Endothermic, energy is absorbed

The correct answer is (D). The reverse of the exothermic reaction in the question is the decomposition reaction, and the reverse of an exothermic process is endothermic. Endothermic reactions absorb energy from the surroundings.

28. If 22.99 g of sodium and 35.45 g of chlorine gas are combined in a vessel and the reaction goes to completion, what will the contents of the vessel contain at the conclusion of the reaction?

(A) Na only

(B) Na, Cl_2, and NaCl

(C) NaCl only

(D) Cl_2 only

The correct answer is (C). The masses given are in the exact, molar stoichiometric ratio given in the balanced equation, meaning each reactant is completely consumed, leaving only the product.

3. **The Free-Response Questions**

The free-response questions fall into two categories: long and short. Three "long" questions should take between 15–20 minutes each to answer, and four "short" questions should take between 7–10 minutes each to answer, but you are free to divide your time up as you see fit. All seven of the free-response questions have multiple parts.

Currently, there is no history of past exam questions related to the new AP Chemistry curriculum, but expect to see an emphasis on science practices and laboratory situations, and on particulate diagrams and models used to represent reactions and concepts at the molecular level.

4. **How Is the Test Scored?**

Section I. Scores on the multiple-choice section of the exam will be based on the number of questions answered correctly. Points are *not* deducted for incorrect answers, and no points will be awarded for unanswered questions.

Section II. Scores on the free-response section are weighted and then combined with the multiple-choice score to give a raw score.

The raw score is converted to a composite score of 5, 4, 3, 2, or 1. An AP score of 5 is approximately equivalent to the average score of a college student who gets an A in an equivalent college chemistry course. A score of 4 is approximately equivalent to college grades of A–, B+, and B, and a score of 3 is approximately equivalent to college grades of B–, C+, and C.

The College Board describes the AP grades as follows:

5 Extremely Well Qualified

4 Well Qualified

3 Qualified

2 Possibly Qualified

1 No Recommendation

5. What Is the Breakdown of AP Chemistry Grades?

	% Students Earning Examination Grade of				
Year	5	4	3	2	1
2010	17.1%	18.5%	19.3%	12.7%	32.3%
2011	17.0%	18.4%	19.5%	14.6%	30.4%
2012	16.4%	19.3%	20.1%	15.0%	29.2%
2013	18.9%	21.5%	18.8%	14.9%	26.0%

The data above indicates that typically about one out of every six students who takes the AP Chemistry examination earns a 5, while staggeringly, almost one out of every three students earns a 1. This *Crash Course* is tailored to help you earn the highest grade possible on the exam.

6. Using Supplementary Materials with Your *Crash Course*

This *Crash Course* contains everything you need to know to score well on the AP exam. You should, however, supplement it with materials provided by the College Board and from your teacher and class. The AP Chemistry Course Description Booklet, and released AP Chemistry exams, can all be found online. Although the format of the AP Chemistry exam was changed in 2014, reviewing previous AP exam questions will still be helpful. The exam has changed, but the chemistry has not.

Note: There are some topics in the AP Chemistry curriculum that are new to the AP Chemistry course as of 2014, and, as such, you will definitely *not* find those topics on old exams. These topics are included in this book and include such concepts as photoelectron spectroscopy, mass spectrometry, an emphasis on particulate diagrams (models), and inquiry laboratory and science practices. There are also some smaller references to other new areas such as capillary action, surface tension, semi-conductors, biological applications, and work.

In addition to that material, our *AP Chemistry All Access* Book + Web + Mobile study system further enhances your exam preparation by offering a comprehensive review book plus a suite of online assessments (chapter quizzes, mini-tests, a full-length practice test, and e-flashcards), all designed to pinpoint your strengths and weaknesses and help focus your study for the exam.

PART II

ATOMS AND ELEMENTS

Chapter 2

Atoms and Moles

I. Atoms

A. Atoms, Compounds, Ratios, and Masses

1. All matter is composed of atoms.
2. There are a limited number of different types of atoms; each different type of atom is an atom of a different element. The elements are organized on the periodic table.
3. A large number of atoms of a single element always has the same average mass.
4. Atoms can combine to form compounds.
5. A pure compound always has the same whole-number ratio of different atoms, e.g., water is always H_2O, and because of #3 above, the ratio of the masses of the elements is always the same in any given compound.
6. If the whole-number ratio of the atoms (and, therefore, the ratio of the masses) changes, then a new compound has been formed. For example, H_2O_2 is *not* water; it is hydrogen peroxide, a different compound with a different ratio of atoms and a different ratio of masses of hydrogen and oxygen atoms.

B. Percentage by Mass and Empirical Formula

1. Calculation of percentage by mass composition. To determine the percentage by mass composition of an individual element within a compound, express the mass of each element as a percentage of the total mass of the compound.

2. For example:
   i. A compound with the formula $C_2H_4O_2$ has a total mass of
   $$(2)(12.01) + (4)(1.008) + (2)(16.00) = 60.05$$
   ii. Percentages of each element expressed as a function of the total mass are shown as follows:

   Carbon $\frac{[(2)(12.01)]}{60.05} = 40.00\%$

   Hydrogen $\frac{[(4)(1.008)]}{60.05} = 6.714\%$

   Oxygen $\frac{[(2)(16.00)]}{60.05} = 53.29\%$

3. All pure compounds have a fixed percentage by mass of each element, so the purity of any given sample can be determined by comparing the percentages by mass. A sample that does not match the percentage by mass of a pure sample is not pure.
4. The empirical formula of a compound is the simplest whole-number (integer) ratio of the atoms of each element in that compound.
5. The empirical formula can be calculated from percentage by mass data.
   i. Take the percentage of each element and assume a sample of 100 g. (This assumption converts percentages to masses.)
   ii. Convert the masses (per 100 g) of each element to the moles (see **II. Moles** below) of each element by dividing each element's mass (per 100 g) by the corresponding average atomic mass taken from the periodic table.
   iii. Find the ratio of the moles calculated in ii above by dividing each of the moles by the smallest number of moles.
   iv. The results from iii above will be in a convenient ratio and gives the empirical formula.

Test Tip

***It may be that the ratio includes a decimal (fraction) such as 0.500, 0.333, 0.250, and so on. Since empirical formulae are the simplest** whole-number ratios, you must multiply all of the numbers in the ratio by 2, 3, or 4 as appropriate in order to remove the decimal (fraction).*

C. Molecular Formula

1. Unlike an empirical formula (which shows the *simplest* whole-number ratio), the molecular formula of a compound shows the *exact* whole-number ratio of the different elements in a compound.
2. Like the empirical formula, the numbers of each element are recorded using a subscript to the right of the element's symbol. When only 1 atom is present, the subscript 1 is assumed (understood) and not written.
3. The following describes the relationship between the molecular formula and the empirical formula.
   i. The molecular formula will be some simple, integer-multiple of the empirical formula. For example, a compound with an empirical formula of CH_2O will have a molecular formula of either CH_2O, or $C_2H_4O_2$, or $C_3H_6O_3$ and so on, where the empirical formula is multiplied by either, 1, 2, or 3, and so on.
   ii. To establish the correct multiplier, and therefore find the molecular formula, it is necessary to know the molecular mass of the compound.
   iii. In the preceding example of an empirical formula of CH_2O, given a molecular mass of 60, divide the molecular mass of the compound by the mass of the empirical formula. In this case, the mass of the empirical formula CH_2O is

$$12.01 + (2)(1.008) + 16.00 = 30.03$$

$$\frac{\text{molar mass}}{\text{mass of empirical formula}} = \frac{60.0}{30.0} = 2$$

iv. The answer in iii is the multiplier, so $(2)(CH_2O) = C_2H_4O_2$, which is the molecular formula.

v. When two entirely different compounds have the same percentage by mass, they will have identical empirical formulas but because they are different compounds, they will have different molecular formulas.

II. Moles

A. Avogadro's Number

1. The mole is a unit of counting used in chemistry. Avogadro's number (6.022×10^{23}) represents the number of particles (atoms, ions, formula units, or molecules) in one mole of any substance.
2. Any atom, element, or compound can have its mass expressed in atomic mass units (amu). The average atomic mass (in amu) for atoms of any element can be found on the periodic table. Atomic mass units can be summed to find the corresponding mass (again in amu) of any given multi-atom species.
3. Atomic mass units are crucially important since the numerical value of the amu of the atoms of a given element is equal to the mass in grams of one mole of that element. For example, a single sodium atom has an average mass of 22.99 amu (taken from the periodic table) and one mole (6.022×10^{23}) of sodium atoms has an average mass of 22.99 grams.
4. This allows the conversion of the number of particles (atoms, ions, formula units, or molecules), the number of moles, and the mass of any given substance. For example:
   - A single (one) formula unit of sodium chloride, NaCl, has a mass of $(22.99 + 35.45) = 58.44$ amu. One formula unit of sodium chloride contains one sodium ion and one chloride ion.

- One mole of sodium chloride has a mass of (22.99 + 35.45) = 58.44 g, which in turn contains one mole (6.022×10^{23}) of sodium ions and one mole (6.022×10^{23}) of chloride ions.
- Similarly, a single molecule of water, H_2O, has a mass of (2)(1.008) + 16.00 = 18.02 amu. One molecule of water contains two hydrogen atoms and one oxygen atom.
- One mole of water has a mass of (2)(1.008) + 16.00 = 18.02 g, which in turn contains two moles $(2)(6.022 \times 10^{23})$ of hydrogen atoms and one mole (6.022×10^{23}) of oxygen atoms.

5. The use of moles is arguably the most important concept in all of chemistry. Balanced chemical equations show the molar ratio of reactant and product particles involved in a chemical reaction, and therefore allow for quantitative relationships (moles and masses) to be determined.

Practice Questions

1. Calculate the empirical formula of a compound containing 40.0% carbon, 6.71% hydrogen, and 53.29% oxygen by mass.

2. Which of the following contains the same number of atoms as 4.032 g of hydrogen atoms?
   (A) 1 mole of H_2
   (B) 6.022×10^{23} atoms of H
   (C) 6.022×10^{23} molecules of H_2
   (D) $(2)(6.022 \times 10^{23})$ molecules of H_2

Answers

1.

	C	H	O
Percentage (%) (and by Assuming a 100 g Sample) Also the Mass in Grams	40.0	6.71	53.29
Molar Mass of the Element	12.01	1.008	16.00
$\text{Moles} = \dfrac{\text{Mass in Grams}}{\text{Molar Mass}}$	3.33	6.66	3.33
Divide by Smallest Number of Moles	$\dfrac{3.33}{3.33} = 1$	$\dfrac{6.66}{3.33} = 2$	$\dfrac{3.33}{3.33} = 1$
Ratio is the Empirical Formula	$C_1H_2O_1$ or CH_2O		

Test Tip

When performing empirical formula calculations, avoid rounding up or down too much in the middle of the calculation, and be lenient with significant figures.

2. **(D)** $(2)(6.022 \times 10^{23})$ molecules of H_2

4.032 g of hydrogen atoms contains 4 moles of hydrogen atoms, as does 2 times 1 mole of hydrogen molecules.

Chapter 3

Electrons

I. Electronic Configuration

A. Schrödinger, de Broglie, Heisenberg, and Coulomb's Law

1. The force of attraction between the negative electrons and the positive nucleus is the basis for the structure of the atom. The force of attraction (or repulsion when the charges are the same) is governed by Coulomb's law, where q_1 and q_2 are the charges, and r is the distance between those charges.

$$F = \frac{q_1 q_2}{r^2}$$

Coulomb's law, and the size of the attractive force that it predicts, determines the ionization energy of any given electron. The ionization energy is the energy required to remove the least tightly held electron in a given species, and it is proportional to the nuclear charge (the number of protons) and the distance between the nucleus and the electron to be removed.

2. Following Bohr's development of the atomic model, the wavelike properties of electrons were incorporated (via quantum mechanics) by Schrödinger, de Broglie, and Heisenberg. This allowed predictions to be made about the specific positions of the electrons within the atom.

3. Electrons are found within very specific, quantized, three-dimensional spaces (called *orbitals*) around the atom, and these spaces are defined by wave functions that are mathematical solutions to the Schrödinger equation.

4. Each three-dimensional space is a probability map where one might expect to find an electron and can be thought of as a "cloud."
5. The *Heisenberg uncertainty principle* states that the position and momentum of an electron can never be simultaneously, exactly known.
6. Each of these three-dimensional spaces (orbitals) is located at a particular distance (or level/shell) from the nucleus. The levels/shells have increasing energy as one moves away from the nucleus.
7. The electronic structure of an atom can be described by the use of shells, sub-shells, orbitals, and spin, which, when taken together, describe the position of any given electron in any given atom.
   i. Each level/shell has a number. The first shell has a principal quantum number (n) = 1, the second n = 2, and so on. In each shell, the maximum number of electrons is given by $2(n^2)$. Using this simple relationship, we can find the maximum number of electrons in each of the first four shells.

Shell Number (n)	Maximum Number of Electrons
1	2
2	8
3	18
4	32

   ii. Each shell is further divided into sub-shells. The number of sub-shells within any level is equal to n, and the sub-shells are given the letters s, p, d, and f. The different types of sub-shells have different three-dimensional shapes; *s* orbitals are spherical, p orbitals are dumb-bell ("figure-eight") shaped and align themselves on *x*, *y*, and *z* axes), and d and f orbitals have more complicated shapes.

Shell Number (n)	Sub-shells
1	s
2	s, p
3	s, p, d
4	s, p, d, f

iii. Each sub-shell is further divided into individual orbitals. Each orbital can hold a maximum of two electrons. The number of orbitals that are possible in each sub-shell, and hence the maximum number of electrons in that sub-shell, is shown below.

Sub-shell	Number of Possible Orbitals	Maximum Number of Electrons
s	1	2
p	3	6
d	5	10
f	7	14

iv. The *Pauli exclusion principle* states that we must be able to distinguish between all electrons in any atom, so any two electrons that are found in the same orbital, in the same sub-shell, and in the same shell, are given different "spins." The "spins" have a value of +½ or –½, thus allowing two electrons that are otherwise identical in terms of their location, to be distinguished from one another.

v. When a species has unpaired electrons present, it is said to be paramagnetic and it will be attracted by an externally applied magnetic field. Those species with only paired electrons are not attracted by the external magnetic field and are called diamagnetic.

vi. Following is a summary of shell, sub-shell, and orbital designation.

Shell Number (n)	Sub-shell Designation	Number of Orbitals in Sub-shell and the Orbital Name	Maximum Number of Electrons in Sub-shell	Maximum Number of Electrons in Shell
1	1s	1 (1s)	2	2
2	2s	1 (2s)	2	8
	2p	3 ($2p_x$, $2p_y$, $2p_z$)	6	
3	3s	1 (3s)	2	18
	3p	3 ($3p_x$, $3p_y$, $3p_z$)	6	
	3d	5 ($3d_{xy}$, $3d_{xz}$, $3d_{x^2-y^2}$, $3d_{yz}$, $3d_{z^2}$)	10	
4	4s	1 (4s)	2	32
	4p	3 ($4p_x$, $4p_y$, $4p_z$)	6	
	4d	5 ($4d_{xy}$, $4d_{xz}$, $4d_{x^2-y^2}$, $4d_{yz}$, $4d_{z^2}$)	10	
	4f	7 (various names)	14	

B. Rules for Filling Orbitals (Aufbau Principle)

1. Determine the number of electrons by referring to the atomic number of the atom.

2. Consider any charges caused by the loss or gain of electrons (i.e., are we dealing with ions?).
    - i. Elements and ions can be isoelectronic with one another, meaning that they have the same electronic configurations; for example, Mg^{2+} and Ne (both have 10 electrons) and S^{2-} and Ar (both have 18 electrons).
3. Lowest energy orbitals are filled first.
4. Orbitals have increasing energies, with 1s having the lowest energy, 2s the next, and so on.
    - i. There is a minor complication here. The 4s orbital has a slightly lower energy than the 3d orbitals and, as a result, the 4s orbital is filled before the 3d orbitals. Similarly, the 5s orbital has a slightly lower energy than the 4d orbitals and, as a result, the 5s orbital is filled before the 4d orbitals. So, when filling the d orbitals, subtract one from the principal quantum number (n) to determine the correct shell.
5. Determining electronic configuration using the periodic table
    - i. The period number and block letter show the shell number and type of electron, respectively.

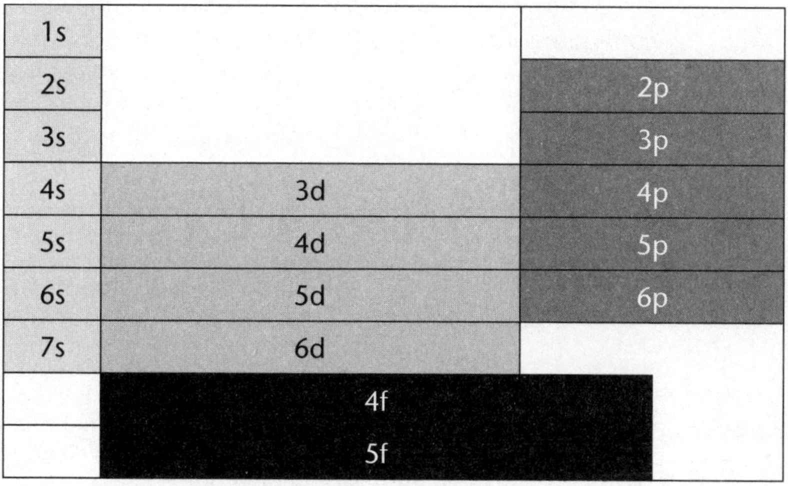

s block—Group 1 and 2
p block—Groups 13 through 18
d block—transition metals
f block—lanthanides and actinides

Test Tip

By far, the easiest way to remember the sequence of orbital filling is by using the periodic table. Label the rows in the table with period numbers and the blocks with the letters s. p, d, and f. Remember to subtract 1 from the period number when entering the d block, subtract 2 from the period number when entering the f block, and re-establish the period number when entering the p block.

ii. Add one electron for each element until the orbital, then the sub-shell, and ultimately the whole shell, is full.

iii. Record the electronic configuration in the format of shell number, type of orbital, and number of electrons (as a superscript). For example:

- Hydrogen has one electron that is found in the s orbital in the first shell, therefore, $1s^1$ (pronounced "one s one").
- Helium has 2 electrons that are found in the s orbital in the first shell, therefore, $1s^2$ (pronounced "one s two").

iv. Other examples: Starting at hydrogen (element #1) each time, follow the periodic table to "build up" the electronic configurations shown in the following table.

Element	# of Electrons	Electronic Configuration
F	9	$1s^2\ 2s^2\ 2p^5$
P	15	$1s^2\ 2s^2\ 2p^6\ 3s^2\ 3p^3$
Sb	51	$1s^2\ 2s^2\ 2p^6\ 3s^2\ 3p^6\ 4s^2\ 3d^{10}\ 4p^6\ 5s^2\ 4d^{10}\ 5p^3$

Test Tip

By adding the superscripts, you can calculate the total number of electrons in any species, given electronic configuration. This can serve as a useful check of your work.

6. The noble gas core method is used to abbreviate electronic configurations. Write the previous noble gas in square brackets and then fill orbitals as before; for example, because the noble gas prior to phosphorus in the periodic table is neon, its electronic configuration can be written as [Ne] $3s^2$ $3p^3$.
7. The orbital diagram notation uses boxes to represent the orbitals and arrows to represent electrons. For example, see the following figure.

Element	Electron Configuration	Orbital Diagram
H	$1s^1$	1s [↑]
He	$1s^2$	1s [↑↓]
Li	$1s^22s^1$	1s [↑↓] 2s [↑]
Be	$1s^22s^2$	1s [↑↓] 2s [↑↓]
B	$1s^22s^22p^1$	1s [↑↓] 2s [↑↓] 2p [↑][ ][ ]
C	$1s^22s^22p^2$	1s [↑↓] 2s [↑↓] 2p [↑][↑][ ]
N	$1s^22s^22p^3$	1s [↑↓] 2s [↑↓] 2p [↑][↑][↑]

   i. The Pauli exclusion principle is illustrated with electrons in the same orbital shown with opposite spins.

8. As can be seen in the preceding figure, the three 2p orbitals avoid having electrons paired until it is absolutely necessary. This is called *Hund's rule of maximum multiplicity*, and it states that if there is more than one orbital with the same energy (called degenerate orbitals), then 1 electron is placed into each orbital before any pairing takes place. All similar orbitals have a similar energy. For example, all three 2p orbitals have the same energy and all five 3d have the same energy. The same is true of the seven f orbitals.
9. Depending on the circumstances, and what one is trying to illustrate, electronic configurations can be represented in a number of different ways. For example, nitrogen (7 electrons) can be represented as follows, each one showing different degrees of detail and having a different emphasis:
   i. $1s^2\ 2s^2\ 2p^3$
   ii. $1s^2\ 2s^2\ 2p_x^1\ 2p_y^1\ 2p_z^1$
   iii. $[He]\ 2s^2\ 2p^3$
   iv. $[He]\ 2s^2\ 2p_x^1\ 2p_y^1\ 2p_z^1$
10. An excited state is said to exist when an electron is promoted to a higher energy level than one might otherwise expect. The expected electronic configuration (lowest energy) is described as the ground state. For example, Mg in the ground state is $1s^2\ 2s^2\ 2p^6\ 3s^2$; in an excited state, it could be $1s^2\ 2s^2\ 2p^6\ 3s^1\ 3p^1$. Many excited states are possible, depending on the extent to which ground state electrons absorb energy.

C. Photoelectron Spectroscopy (PES)—Evidence for the Shell Model and Orbitals

   1. High energy X-rays or UV photons can be used to eject electrons from atoms. This is called the photoelectric effect. The energy applied can be calculated using

$$E = h\upsilon$$

   where h = Planck's constant and υ = frequency.

2. The electrons that are ejected can be analyzed using PES to produce a spectrum that shows peaks that correspond to the energy (*x*-axis) and the relative number of electrons in any given sub-level (*y*-axis).
3. Electrons that are close to the nucleus, therefore with a greater attraction for the nucleus based upon Coulomb's law, will require larger energies to eject them.
4. Analysis of PES data leads to evidence for the shell model. For example, knowing the electronic configuration of sodium to be $1s^2\ 2s^2\ 2p^6\ 3s^1$, we would expect;
   i. Four peaks on the PES plot corresponding to the four different energies required to remove electrons from different sub-shells, with increasing energies as one approaches the nucleus, and
   ii. Peaks with heights (intensities) relative to the number of electrons in each sub-shell, i.e., for Na in the ratio, 2:2:6:1.

   A simulated PES spectrum for sodium is shown below. Note that the scale on the *x*-axis is not linear, that electrons in different sub-shells but the *same* shells tend to be close together in terms of energy, and that electrons in *different* shells are often widely separated in terms of energy. Also note that the relative size (height) of each peak corresponds to the relative number of electrons in each sub-shell.

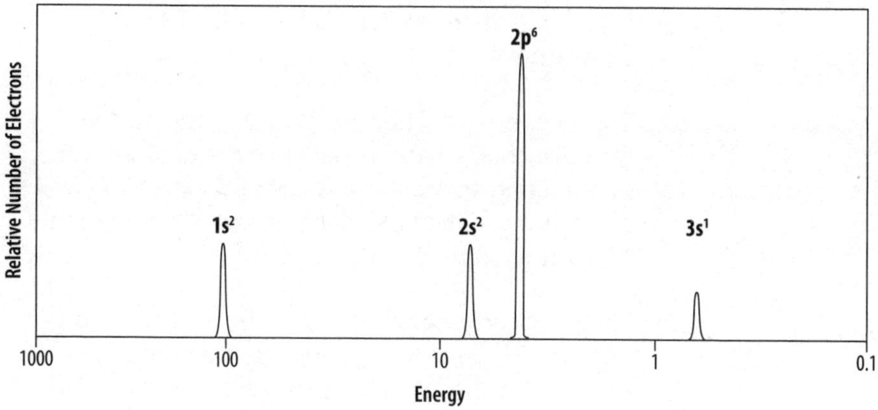

Practice Questions

1. Aluminum has an electronic configuration of $1s^2\ 2s^2\ 2p^6\ 3s^2\ 3p^1$.
   (a) How many unique peaks are expected in aluminum's PES spectrum? Explain.
   (b) Which electrons correspond to the largest energies? Explain.
   (c) Which peak in the spectrum will have the greatest intensity (i.e., be the largest)? Explain your answer.
   (d) Explain why the peaks corresponding to the 3s and the 3p electrons are relatively close together, and why they are distinctly different to the electrons in the 2s and 2p orbitals.

2. An element in the third period of the periodic table produces a PES spectrum that has only three peaks, in the ratio 2:2:1. Identify the element.
   (A) Aluminum
   (B) Boron
   (C) Carbon
   (D) Sodium

Answers

1. (a) Five, since there are five distinct sub-shells present in an Al atom.

(b) The 1s electrons. They are closest to the nucleus (positive charge) and experience the greatest force of attraction according to Coulomb's law. They will be the most difficult to remove and will require the greatest energy.

(c) The peak corresponding to the 2p electrons. The intensity of the peak is relative to the number of electrons in the orbital. The 6, 2p electrons create a peak with the greatest intensity.

(d) Even though the 3s and 3p electrons are in distinct orbitals that correspond to different energies, they are both in the third shell, at similar distances from the nucleus, and thus have similar energies. The 2s and 2p electrons on the other hand are a whole shell closer to the nucleus, and as such are much more difficult to remove and require distinctly higher energies.

2. **(B)** Boron

Boron has an electronic configuration of $1s^2\ 2s^2\ 2p^1$ that would produce a PES spectrum with three peaks in the ratio 2:2:1.

Chapter 4

Periodicity

I. The Periodic Table

A. Periodicity

1. The elements that make up the periodic table are arranged in order of increasing atomic number and in a pattern that is consistent with regular changes of electronic configuration. Vertical columns are called groups; horizontal rows are called periods.
2. With regular electronic configuration patterns come regular patterns of atomic properties.
3. Periodicity can be used to predict the properties of substances by comparing them to similar, known substances.

B. Periodic Relationships

1. Atomic size. As a period is traversed from left to right, the atomic size decreases. This is because the nuclear charge increases (greater positive charge, extra protons) but the subsequent, extra electrons enter the same quantum level, experiencing no extra shielding (see 4. iv. below) from inner electrons and are therefore attracted (pulled in) more tightly.

   As a group is descended, the atomic size increases. Although there is an increase in the number of protons, there is also an increase in quantum levels, so the extra electrons are further away from the nucleus and experience more shielding (see 4. iv. below). As a result, the atomic size increases because the greater the number of quantum levels occupied in an atom, the larger the atom.

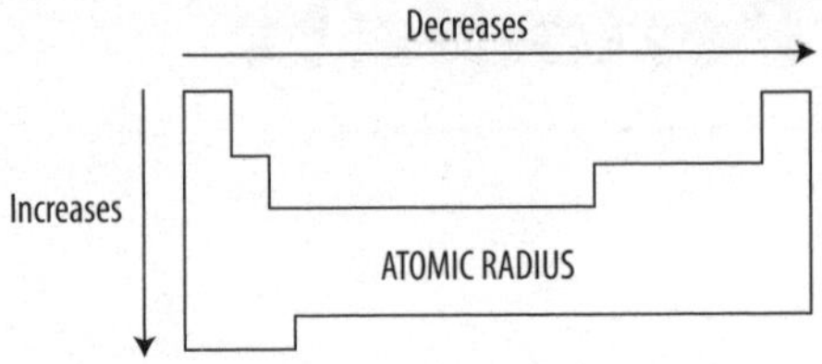

2. Cation (positive ion) size. When an atom loses electrons to form a cation, the remaining electrons will experience less mutual repulsion. As a result, they are drawn in closer than in the parent atom, and, therefore, the cation is smaller than the parent atom. It is also true that when a cation is formed, an atom often loses a complete valence shell of electrons and that the effective nuclear charge (see 4. iv. below) increases, since there are less electrons shielding the nuclear charge. All have the effect of decreasing the size of the cationic species compared to the parent atom.
3. Anion (negative ion) size. When an atom gains electrons to form an anion, the electrons that have been added to form the anion repel one another, and the increased mutual repulsion makes the anion larger than the parent atom. It is also true that when an anion is formed that the effective nuclear charge (see 4. iv. below) decreases, since there are more electrons shielding the nuclear charge. All have the effect of increasing the size of the anionic species compared to the parent atom.
4. Ionization energy. Ionization energy is the energy required to remove electrons.
   i. *First ionization energy* is defined as the energy required to remove 1 mole of electrons from 1 mole of gaseous atoms to produce 1 mole of gaseous ions.

$$M_{(g)} \rightarrow M^{+}_{(g)} + e^{-}$$

ii. *Second ionization energy* is defined as the energy change accompanying the process shown below.

$$M^{+}_{(g)} \rightarrow M^{2+}_{(g)} + e^{-}$$

iii. Ionization energies are measured in units of kJ mol^{-1}. They have positive (endothermic) values, indicating that energy must be "put in" in order to remove electrons. Energy is required, since there is a Coulombic force of attraction between the positive nucleus and the negative electrons.

iv. The following are factors that affect the magnitude of the ionization energy. The attraction of electrons to the nucleus is dependent upon:

- The nuclear charge (how many protons are present).
- The shielding effect of the inner electrons (the extent to which inner electrons protect the outer electrons from the nuclear charge).

These factors are sometimes combined and expressed in a term called *effective nuclear charge* or Z_{eff}. It is calculated by subtracting the shielding effect of the electrons from the positive charge of the nucleus.

v. Large jumps in successive ionization energies for a single element are observed when passing from one quantum shell to another. In the following example, element X loses three electrons relatively easily, but the fourth requires much more energy. This is because it is situated in a new quantum shell, closer to the nucleus with less shielding. As a result, we can predict X has three electrons in its outer shell and therefore is in group 13 of the periodic table.

Ionization Energies for Element X (kJ/mol)				
First	Second	Third	Fourth	Fifth
540	1651	2650	14921	17345

vi. The following are periodic trends in ionization energy:

- When crossing a period from left to right, the ionization energy will generally increase. This is because the nuclear charge increases (greater positive charge, extra protons) and the electrons are being removed from the same principal quantum level (shell), experiencing no extra shielding, and therefore are held more strongly.
- When descending a group from top to bottom, the ionization energy will decrease because, although there is an increase in nuclear charge, the outer electrons enter new quantum levels (shells) further away from the nucleus and experience more shielding from the inner electrons and therefore are held less strongly. Shielding is the more important factor.

5. Electron affinity. Electron affinity is the quantitative measurement of energy changes that occur when adding electrons to atoms or ions.

i. *First electron affinity* is defined as the energy change when 1 mole of gaseous atoms gains 1 mole of electrons to form 1 mole of gaseous ions

$$X_{(g)} + e^- \rightarrow X^-_{(g)}$$

ii. *Second electron affinity* is defined as the energy change accompanying the process shown below.

$$X^-_{(g)} + e^- \rightarrow X^{2-}_{(g)}$$

iii. Electron affinities are measured in units of kJ mol^{-1}. They have both positive (endothermic) and negative (exothermic) values depending on the species being formed. This indicates that energy may have to be "put in" (positive) or energy may be "released" (negative) when adding electrons.

iv. Patterns of electron affinity are less easy to predict than patterns of ionization energy; however, it is useful to know the following:

- The overall trend from left to right on the periodic table is an increasing tendency to accept electrons, i.e., that electron affinities become increasingly positive. This makes sense because nonmetals (on the right side of the periodic table) tend to want to form negative ions (by accepting electrons) much more readily than metals (on the left side of the table) that generally tend to want to lose electrons and form positive ions. As a result, group 17 (the halogens) have the highest electron affinity values.
- The overall trend within a group is a little more difficult to predict, but as a general rule, values vary little among elements of the same group but with some small increases as one passes from the bottom to the top of a group.

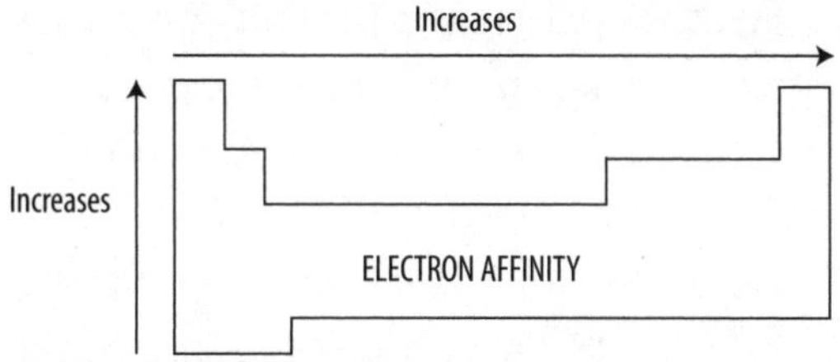

6. Electronegativity. Electronegativity is a measure of how well atoms attract shared electrons.
   i. Electronegativity is defined as the ability of an atom to attract electrons to itself within a covalent bond.
   ii. Electronegativity increases across a period from left to right and increases up a group from top to bottom.
   iii. Fluorine is the most electronegative element.
   iv. The noble gases (group 18) are usually omitted from electronegativity discussions because they form so few covalent bonds.

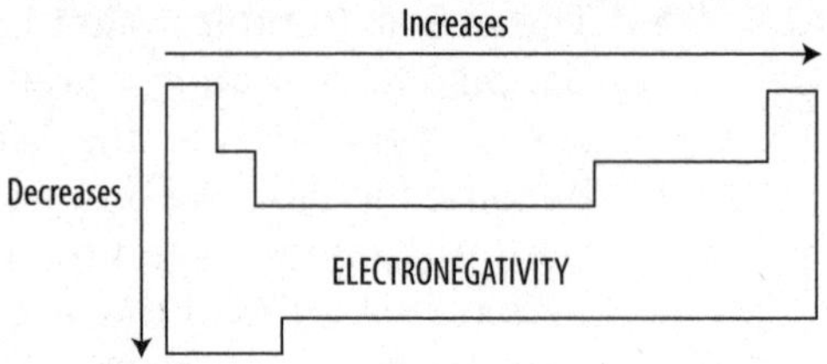

7. Typical ionic charges

   i. Group 1—Alkali Metals (Li, Na, K, Rb, Cs, Fr)

      ➤ Tend to lose 1 valence electron to become stable (noble gas configuration) and, therefore, have a +1 charge

   ii. Group 2—Alkaline Earth Metals (Be, Mg, Ca, Sr, Ba, Ra)

      ➤ Tend to lose 2 valence electrons to become stable (noble gas configuration) and, therefore, have a +2 charge.

   iii. Group 13—The Boron Group (B, Al, Ga, In, Tl)

      ➤ Tend to lose 3 valence electrons to become stable (noble gas configuration) and, therefore, have a +3 charge.

   iv. Group 14—The Carbon Group (C, Si, Ge, Sn, Pb)

      ➤ A mixture of losing, gaining, and sharing electrons to become either cations, anions, or involved in covalent bonding respectively

      ➤ Tin (Sn^{2+}, Sn^{4+}) and lead (Pb^{2+}, Pb^{4+}) common ions, C and Si commonly found in covalent compounds

   v. Group 15—The Nitrogen Group (a.k.a. Pnictogens) (N, P, As, Sb, Bi)

      ➤ Tend to gain 3 valence electrons to become stable (noble gas configuration) and, therefore, have a –3 charge.

   vi. Group 16—The Oxygen Group (a.k.a. Chalcogens) (O, S, Se, Te, Po)

      ➤ Tend to gain 2 valence electrons to become stable (noble gas configuration) and, therefore, have a –2 charge.

vii. Group 17—Halogens or "salt formers" (F, Cl, Br, I, At)

➤ Tend to gain 1 valence electron to become stable (noble gas configuration) and, therefore, have a –1 charge.

viii. Group 18—Noble Gases (He, Ne, Ar, Kr, Xe, Rn)

➤ Have a full outer s and p sub-shells of 8 electrons and, therefore, are stable and tend not to form ions.

ix. Transition Metals (*d*-block elements)

➤ Lose electrons to have multiple positive charges based on what provides for the most stable electronic configurations

➤ Lose s electrons before d electrons when forming ions

8. Periodic physical properties of the elements. There is variation in physical properties in groups and periods.

   i. Physical properties within a group. The value of a property tends to change relatively uniformly from top to bottom in a group. For example, consider some of the physical properties of the elements of Group 17.

Selected Physical Properties of Group 17 Elements

	Melting Point in Kelvin (K)	Boiling Point in Kelvin (K)	State at Room Temperature	Color
Fluorine	53	85	Gas	Pale green
Chlorine	172	239	Gas	Yellow/green
Bromine	266	332	Liquid	Brown/orange
Iodine	387	458	Solid	Purple/black

ii. Physical properties across a period. Periodic physical properties are less easy to predict with certainty, but sometimes the value of a property may reach a peak within a period and then reverse the trend. For example, consider the melting point of the third-period elements. Here, the melting points rise to a peak at silicon before falling back to smaller values than at the beginning of the period.

Melting Points of Period 3 Elements

Element	Melting Point in K
Na	371
Mg	922
Al	933
Si	1683
P	863
S	393
Cl	172
Ar	84

9. Periodic chemical properties of binary compounds (oxides). There is variation in chemical properties of the oxides.
   i. Metal oxides of groups 1 and 2 are basic and form hydroxides in solution.
   ii. Non-metal oxides in groups 14–17 are acidic and form acids in solution.

Practice Question

1. Based on the data below, element X is in which group of the periodic table?

Ionization Energies for Element X (kJ/mol)

First	Second	Third	Fourth	Fifth
540	1651	2650	14921	17345

(A) Group 1
(B) Group 2
(C) Group 13
(D) Group 14

Answer

1. (C) Group 13.

Since the largest energy gap is from third to fourth, element X must achieve noble gas configuration after losing 3 electrons.

Atomic Models

I. General Principles

A. Atomic Models

1. Since atoms are so incredibly tiny, they are very difficult to study directly, so models can be used to explain their behavior and to interpret any data about atoms that is generated experimentally.
2. As in all science, models and explanations must be refined over time when new evidence or experimental data comes to light. This is crucial when evidence that contradicts or is inconsistent with the current model is found.

B. Mass Spectrometry

1. John Dalton originally stated that all atoms of a given element were identical. Later, new evidence indicated otherwise, and his model of the atom had to be refined.
2. We now know that different atoms of the same element can have varying numbers of neutrons and, as a result, varying atomic masses. This led to Dalton's original hypothesis being modified. Dalton originally stated that all atoms of a given element were identical. We now know that atoms of the same element can have differing numbers of neutrons and, hence, different atomic masses.
3. Atoms of the same element with differing numbers of neutrons are called isotopes.
4. Mass spectrometry (the technique) and the mass spectra (the hard-copy results) that the mass spectrometer (the machine) produces are used in modern chemistry to identify the

varying masses of the isotopes in a naturally occurring sample of any given element.

5. A mass spectrometer works as follows:
   i. Ionization. Positive ions are formed by a sample of vaporized atoms being bombarded with electrons. These electrons knock valence electrons out of the sample atoms, thus creating positive ions in the sample. (Positive ions are always formed, even when dealing with atoms that normally form negative ions.)
   ii. Acceleration. The positive ions are accelerated into a singular beam by the use of an electric field.
   iii. Deflection. The positive ions are deflected in a magnetic field according to their mass and charge (mass/charge ratio). Lighter and more highly charged ions are deflected more than heavier and lower charged ions.
   iv. Detection. The beam of deflected positive ions are detected and converted to a readable output (a mass spectrum).
6. A mass spectrum is a plot of charge/mass ratio on the *x*-axis, compared to the relative abundance or intensity on the *y*-axis. A peak in the spectrum indicates the existence of an isotope with that mass (assuming charges to be +1), and the relative height of the peak is consistent with the relative abundance of that particular isotope. For example, the mass spectrum for elemental chlorine is:

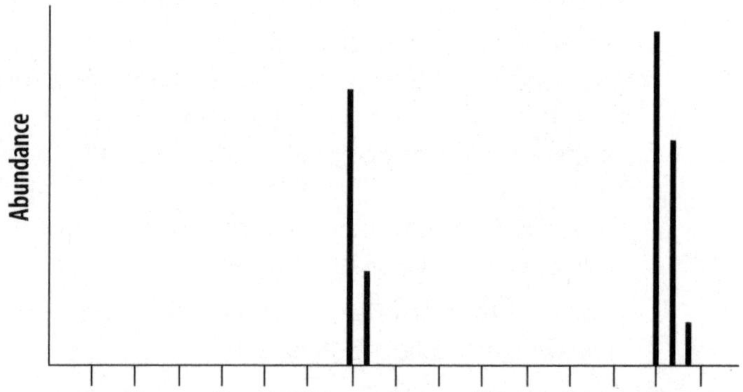

The common diatomic elements (Br, I, N, Cl, H, O, and F) will also exhibit peaks in the mass spectrum that reflect the masses of pairs of atoms that make up the molecule. For example, chlorine has two common isotopes, Cl-35 and Cl-37, and they combine in diatomic molecules to give peaks in the mass spectrum at 70 (35 + 35), 72 (35 + 37), and 74 (37 + 37) as well as the individual isotopic peaks of the atoms at 35 and 37.

7. The mass spectrum of any element can be used to calculate the average atomic mass of that element by using the masses of the isotopes in a weighted average calculation. For example, consider the mass spectrum of elemental molybdenum, below.

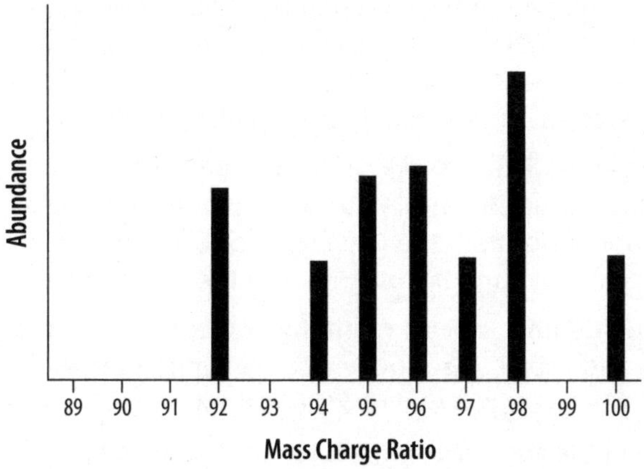

The mass spectrum shows that there are seven isotopes of molybdenum of masses 92, 94, 95, 96, 97, 98, and 100. These isotopes have abundances of 14.84%, 9.25%, 15.92%, 16.68%, 9.55%, 24.13%, and 9.63%, respectively. The average mass of molybdenum (Mo) can be calculated as

$$(92)(.1484) + (94)(.0925) + (95)(.1592) + (96)(.1668) + (97)(.0955) + (98)(.2413) + (100)(.0963) = 96.03$$

Test Tip

It is possible that data for relative abundances might be provided that are not percentages, but are relative numbers. If and when this happens, still divide the sum of the products of the mass and abundance data by the total. In these circumstances, the total will not be 100. For example, if Cl-35 and Cl-37 are found in a ratio 3:1, then the average atomic mass is calculated as (35)(3/4) + (37)(1/4) = 35.5.

C. Using Energy to Investigate Atoms, Molecules, and Concentrations

1. Infrared radiation is used to cause the bending and stretching of covalent bonds. Since different types of covalent bonds will interact with infrared radiation in different ways, IR spectroscopy can be used to identify the presence of different types of covalent bonds present in larger molecules. IR spectroscopy is used mainly in organic chemistry.
2. Ultraviolet radiation and visible light are used to cause electron transitions within atoms (promotion to higher energy levels and ionization energies) and to collect data about the electronic structure of the atom and the electron shell model.
3. The quantitative aspects of interaction between radiation and atoms and molecules is governed by Planck's equation, $E = h\nu$, where E = energy, h = Planck's constant, and ν = frequency.
4. The Beer-Lambert law is used to assess the concentration of a solution (*almost exclusively used in colored solutions on AP exams*), by relating the absorbance of the solution to the concentration of the solution.
   i. The equation used is $A = abc$, where A = absorbance, a = absorptivity coefficient (a constant for a particular solution), b = path length (the distance that the light passes through the solution), and c = concentration.
   ii. A colorimeter (or similar instrument like a Spec 20) is used, and the wavelength of the light chosen for the experiment should be complementary to the wavelength of the light reflected by the solution (i.e., complementary to the color of the solution). This allows for maximum absorbance (an important aspect of collecting reliable data over a range of concentrations).

iii. Linear plots of absorbance versus concentration are found.

iv. The technique is generally only useful for colored solutions.

Practice Questions

1. The mass spectrum of neon shows three peaks at mass/charge ratios of 20, 21, and 22 with abundances of 90.48%, 0.27%, and 9.25%, respectively.
   (a) Calculate the average relative atomic mass of Ne based upon this data.
   (b) Which of the three peaks would be the tallest? Explain.

2. Several solutions of nickel (II) sulfate, ranging from 1.00 M to 0.200 M, are exposed to light in a colorimeter, and their absorbance is recorded in a Beer-Lambert law experiment.
   (a) Why does a solution of nickel (II) sulfate represent a suitable solution to be studied in this experiment, whereas similar solutions of sodium chloride do not?
   (b) What general relationship would one expect to find between absorbance and concentration in the nickel (II) sulfate solutions?
   (c) Nickel (II) sulfate solutions are green. Why would light of a wavelength equivalent to the green part of the visible spectrum be a poor choice for use in this experiment?
   (d) In the equation $A = abc$, what does b represent?
   (e) Explain how the absorbance and concentration data collected in such an experiment can be used to determine the concentration of an unknown solution of nickel (II) sulfate.

Answers

1. (a) (20)(.9048) + (21)(.0027) + (22)(.0925) = 20.19

   (b) The peak corresponding to the isotope with mass of 20, since it is the most abundant. The height of peaks corresponds to relative abundance.

2. (a) A solution of nickel (II) sulfate is green but a solution of sodium chloride is colorless. As such, sodium chloride solutions do not vary in absorbance as a function of concentration.

   (b) Directly proportional. As concentration goes up, colored solutions get darker and absorbance goes up (b and c in the equation $A = abc$ are constant).

   (c) Green solutions reflect green light and will not absorb light with a wavelength in the green part of the visible spectrum (that's what makes those solutions green). In experiments such as this, one must use radiation with a wavelength that causes the maximum absorbance, i.e., the opposite of green light (the complementary color to green light) which is red light.

   (d) b = path length, the distance that the light travels through the solution in the cuvette in the colorimeter.

   (e) A graph of absorbance (y-axis) versus concentration (x-axis) will yield a straight line (a linear relationship). Then, by either reading from the graph, extrapolating, or using the equation for the line, other values of concentration can be determined from the absorbance observed for the unknown solution.

Conservation of Atoms and Mass

I. Conservation of Atoms and Mass

A. The Representation of Atoms

1. Atoms and mass are conserved (neither created nor destroyed) in any chemical or physical process.
2. A chemical equation (symbolic representation) reflects the conservation of atoms and mass by being balanced. For example, in the balanced chemical equations below, all carbon, hydrogen, and oxygen atoms appear in equal numbers on either side of the equations.

$$CH_4 + 2O_2 \rightarrow CO_2 + 2H_2O$$

$$2C_2H_6 + 7O_2 \rightarrow 4CO_2 + 6H_2O$$

$$C_3H_8 + 5O_2 \rightarrow 3CO_2 + 4H_2O$$

3. A particulate representation also reflects the conservation of atoms by showing equal numbers of particles before and after reaction. For example, the first reaction in #2 above can be represented as

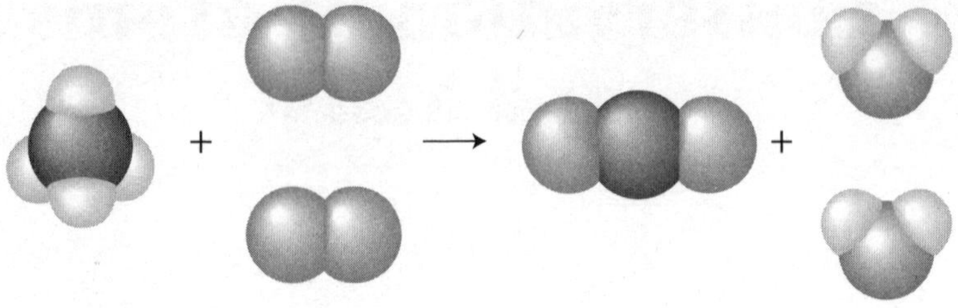

4. If the mass of the reactants is known, because atoms (and therefore mass) are conserved, it is possible to calculate reactant or product masses from product or reactant masses, respectively. For example, in the combustion of methane, $CH_4 + 2O_2 \rightarrow CO_2 + 2H_2O$, using stoichiometric amounts, if 1.6 g of CH_4 react with exactly 6.4 g of O_2 to produce 4.4 g of CO_2, then since the total mass of the reactants is 8.0 g, the missing product mass of H_2O must be 8.0 g – 4.4 g = 3.6 g.
5. The conservation of atoms and mass means that as atoms travel throughout the universe, from one situation to another, they are never created or destroyed; rather they are simply rearranged into different compounds.

B. Balanced Chemical Equations and Moles

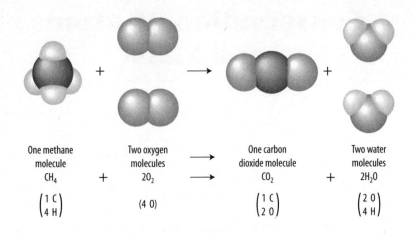

1. Chemical formulas show the ratio of different atoms in any given molecule or compound. For example, H_2O shows two atoms of hydrogen are combined with one atom of oxygen in each molecule of water.
2. Chemical formulas can be converted to masses by using atomic mass data from the periodic table. For example, H_2O has a total mass of (2)(1.008) + (16.00) = 18.02 amu per molecule of water, or 18.02 g per mole of water.
3. Balanced chemical equations bring chemical formulas together to show how they change in chemical reactions. They have stoichiometric coefficients that show the relative

numbers of particles and moles that react. For example, $CH_4 + 2O_2 \rightarrow CO_2 + 2H_2O$ shows that 1 mole (or molecule) of CH_4 reacts with 2 moles (or molecules) of O_2 to form 1 mole (or molecule) of CO_2 and 2 moles (or molecules) of H_2O. In the process, all atoms are conserved.

C. Using Moles in Gravimetric Analysis and Titrations (Volumetric Analysis)

1. Gravimetric analysis uses the formation of a solid and a balanced chemical equation to allow the quantitative analysis of a dissolved analyte. One common example of the technique is the analysis of aqueous chloride ions by the addition of aqueous silver ions.

$$Ag^{+}_{(aq)} + Cl^{-}_{(aq)} \rightarrow AgCl_{(s)}$$

Addition of excess, aqueous silver ions (often in the form of silver nitrate) will cause the chloride ions to precipitate out, forming a solid that can be filtered, dried, and massed. Knowing that mass and atoms must be conserved and that chemical equations show the ratio of the moles, the composition of the original sample (in terms of chloride ions) can be determined.

2. The titration technique uses a titrant, with a known concentration, being added to the analyte up to the equivalence point. The equivalence point is the point at which the analyte is exactly and totally consumed by the titrant, and must be accompanied by some observable change, often when an indicator changes color. The point at which the color change is observed is known as the end-point. One simple example is an acid–base titration between solutions of hydrochloric acid and sodium hydroxide with a methyl orange indicator.

$$HCl_{(aq)} + NaOH_{(aq)} \rightarrow NaCl_{(aq)} + H_2O_{(l)}$$

Knowing the volume (in L) and concentration (in mol/L) of the HCl, the moles of HCl can be calculated. Knowing the volume of the NaOH, that mass and atoms must be conserved, and that chemical equations show the ratio of the moles, the concentration of the NaOH can be determined.

Practice Question

1. An impure solid contains some chloride ions. A 3.00 g sample of this solid is dissolved in water and the chloride ions dissolve to produce an aqueous solution. The solution has a large amount of aqueous silver ions added to it and a white precipitate of silver chloride forms. Silver ions are added until no more precipitate appears. After filtering and drying, it is found that 0.450 g of solid is produced in the reaction. Calculate the mass of chloride ions in the original, impure sample.

Answer

1. Moles of AgCl = (0.450 g/143.32 g mol^{-1}) = 3.14 x 10^{-3} mols

   According to the equation, $Ag^+_{(aq)} + Cl^-_{(aq)} \rightarrow AgCl_{(s)}$, the moles of chloride ions are equal to the moles of silver chloride (1:1 ratio).

   Mass of chloride ions = (3.14 x 10^{-3} mols)(35.45 g mol^{-1}) = 0.111 g

PART III

BONDING

Chapter 7

Solids, Liquids, Gases, and Solutions

I. Solids and Liquids

A. Solids

1. All solids have particles that have limited energies and limited movement, where those particles can pack closely together but only vibrate around fixed positions. The particles do not change positions relative to one another.
2. Solids can be arranged in regular, orderly, 3-D structures, i.e., they are crystalline (as in ionic solids, for example), or they are amorphous where the particles are *not* orderly arranged.

B. Liquids

1. All liquids have particles that are close together and that move around one another and collide with one another.
2. Because of the relative ease of movement, unlike in solids, the particles of a liquid *do* change position relative to one another.

C. Comparisons of Solids and Liquids

1. The solid and liquid phases of a particular substance tend to have similar molar volumes since in both states, the particles tend to be very close together.
2. In all cases, the properties of solids and liquids are wholly dependent upon:
   i. the strength of the attraction between the particles present; and
   ii. the manner in which those particles are arranged

For example, if the particles in a liquid have strong intermolecular forces, then the liquid is likely to be more viscous. The strength of the inter-particulate forces within various solids can affect properties such as hardness.

II. Gases

A. General Properties

1. They are compressible (they have a large amount of space between their component particles).
2. They have no definite shape or volume (they take on the shape and volume of their container).
3. They "fill" their containers (they spread out to occupy the greatest possible space).
4. They occupy more space than solids or liquids.
5. When their particles collide with the walls of their container, they exert a pressure that is measurable.
6. They have relatively low densities when compared to liquids and solids.
7. At STP—Standard Temperature and Pressure (standard temperature is 0°C (273K) and standard pressure is 1 atm = 760 mm Hg = 760 torr), 1 mole of *any* ideal gas occupies a volume of 22.4 L.
8. They must have their quantitative properties calculated using temperatures in Kelvin.

B. Kinetic Molecular Theory (KMT) and Ideal Gases

1. KMT has five postulates:
   i. The KE of a gas is directly proportional to the Kelvin temperature. The higher the temperature, the greater the kinetic energy of the gas particles. Kinetic Energy (*KE*) is defined as the energy of motion.

$$KE = \frac{1}{2}\text{mass} \times \text{velocity}^2$$

ii. Gases consist of particles whose volume is negligible compared to the volume of the container, i.e., the gas particles themselves effectively occupy zero space.

iii. Gas particles are in continuous, random, and rapid motion.

iv. Gas particles collide with each other and the walls of their container. During these collisions, no energy is lost, i.e., they are elastic collisions.

v. Gas particles do not attract one another, i.e., effectively, each gas particle moves independently.

C. Speeds of Gases

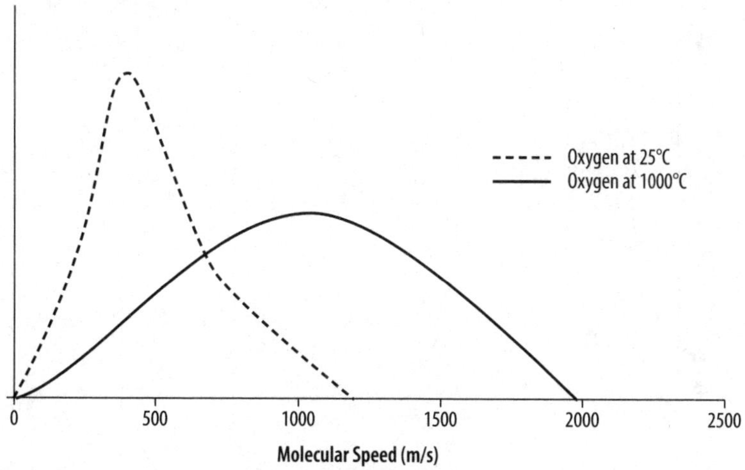

1. The data above indicates the following relating to the speed of gases

   i. At any given temperature, gas molecules have both high and low speeds.

   ii. As temperature increases, the distribution of speeds is found to be across a wider range.

   iii. As the temperature increases, a greater number of molecules are traveling faster. For example, at 1000°C, molecules can be traveling between 1500 and 2000 m/s, while at 25°C they cannot be traveling at speeds in that range.

iv. Root mean square (rms) speed ($\sqrt{u^2}$), temperature (T, in Kelvin), and molar mass (*MM*) are related:

$$\sqrt{u^2} = \sqrt{\frac{3RT}{MM}}$$

On the AP Chemistry exam, you will not be asked to perform calculation of root mean square speed, but you need to appreciate that it is dependent upon molar mass (i.e., that heavier particles move more slowly) and upon temperature (i.e., that particles at higher temperatures move more quickly). You could see a qualitative question based upon these facts.

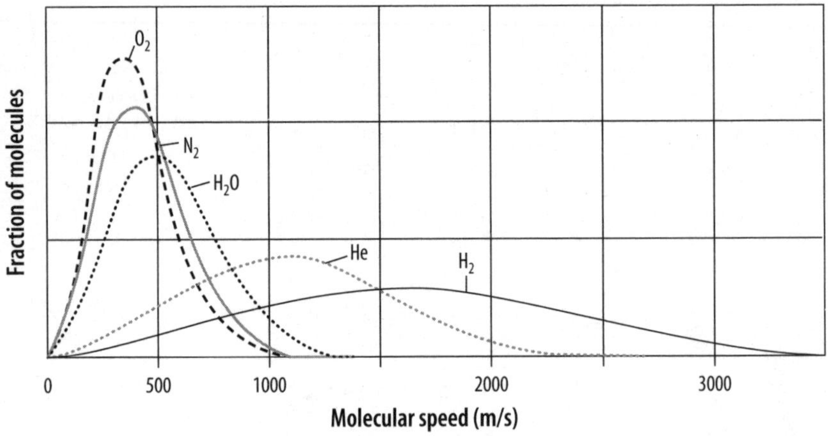

2. The data above indicates that at any given temperature, heavier gas molecules have lower speeds.

D. Mathematical Relationships for Gases

Test Tip

All calculations performed on gases should use temperatures in Kelvin.

1. The ideal gas equation can be used to relate the properties [pressure (P), volume (V), number of moles (n), and temperature (T)] of ideal gases to one another.

$$PV = nRT$$

Where R = universal gas constant = 8.314 J mol^{-1} K^{-1} = 0.08206 L atm mol^{-1} K^{-1} = 62.36 L torr mol^{-1} K^{-1} (i.e., the numerical value of R is dependent upon the units used).

2. Boyle's Law states that at constant temperature and with a constant mass of gas, pressure is inversely proportional to volume.

$$PV = \text{a constant}$$

If the pressure and volume of a gas are known initially and one of the variables is changed, then the new conditions can be calculated by using $P_1V_1 = P_2V_2$. *(P_1 and V_1 are the original conditions, and P_2 and V_2 are the new conditions. Units of pressure and volume must be the same on each side of the equation.)* See the following diagram.

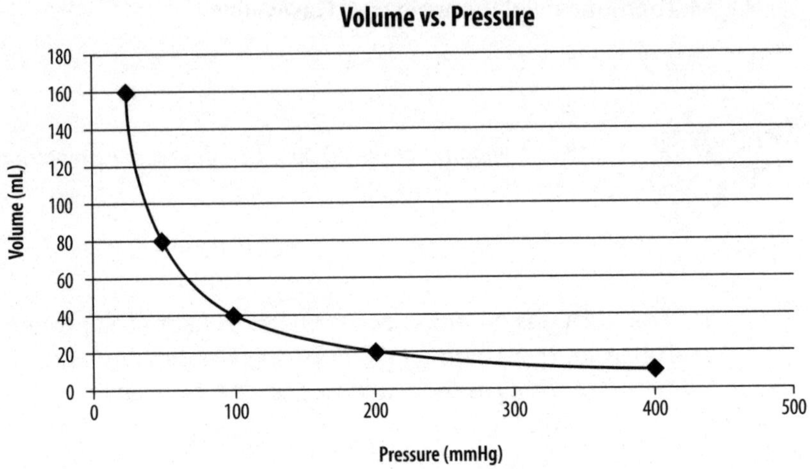

Note in the graph above that when the pressure is doubled, the volume is halved. Therefore, the mathematics of solving a Boyle's Law problem is easy.

3. Charles's Law states that at constant pressure and with a constant mass of gas, volume is directly proportional to the Kelvin temperature.

$$\frac{V}{T} = \text{a constant}$$

If the volume and temperature of a gas are known initially and one of the variables is changed, the new conditions can be calculated by using $\frac{V_1}{T_1} = \frac{V_2}{T_2}$. *($V_1$ and T_1 are the original conditions, and V_2 and T_2 are the new conditions. Units of volume must be the same on each side of the equation, and temperature must be in Kelvin.)* See the following diagram.

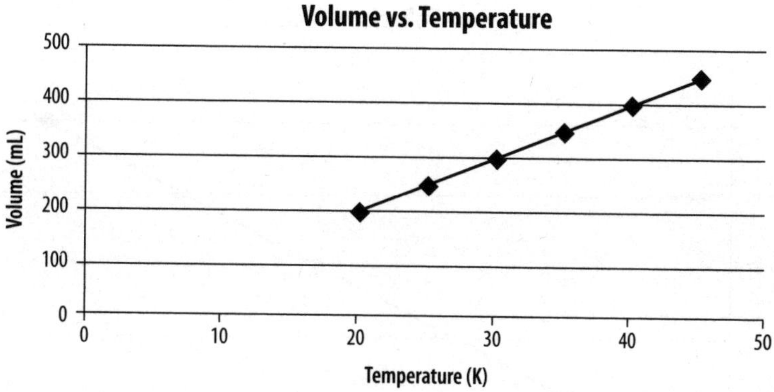

> Note in the graph above that when the temperature is doubled, the volume is doubled. Therefore, the mathematics of solving a Charles's Law problem is easy.

4. Avogadro's Law states that at constant temperature and constant pressure, volume is directly proportional to the number of moles of gas present.

$$\frac{V}{n} = \text{a constant}$$

If the volume and number of moles of a gas in known initially and one of the variables is changed, the new conditions can be calculated by using $\frac{V_1}{n_1} = \frac{V_2}{n_2}$. *($V_1$ and n_1 are the original conditions, and V_2 and n_2 are the new conditions. Units of volume and moles must be the same on each side of the equation.)* See the following diagram.

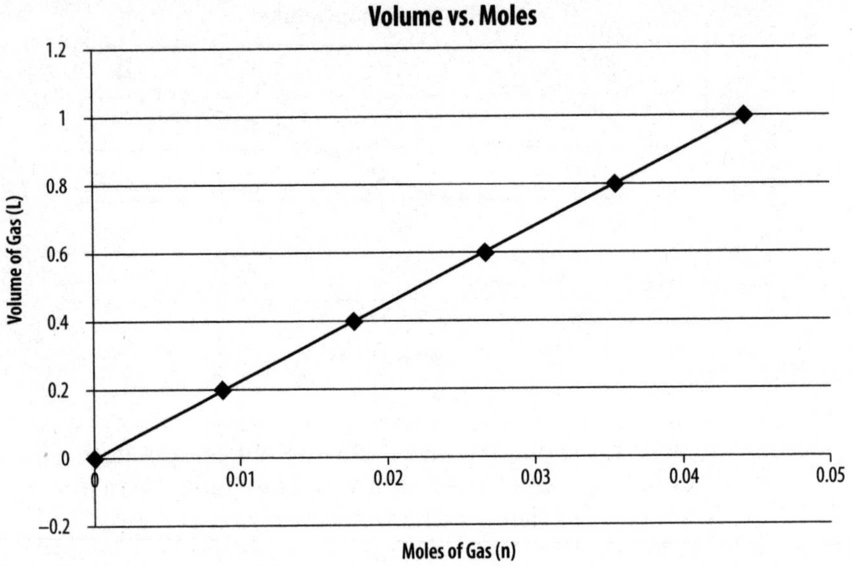

5. Gay-Lussac's Law states that, at constant volume and with a constant mass of gas, pressure is directly proportional to temperature.

$$\frac{P}{T} = \text{a constant}$$

If the pressure and temperature of a gas is known initially and one of the variables is changed, the new conditions can be calculated by using $\frac{P_1}{T_1} = \frac{P_2}{T_2}$. *($P_1$ and T_1 are the original conditions, and P_2 and T_2 are the new conditions. Units of pressure must be the same on each side of the equation, and temperature must be in Kelvin.)*

Test Tip

A very useful strategy in dealing with gas calculations is to make a list of the values that are given in any question and noting the variable that is unknown. Very often this will help you to decide which equation to use. For example, if values for an initial pressure, new pressure, and initial volume are given in the question (i.e., P_1, P_2, and V_1) and the question is asking for the new (final) volume (i.e., V_2 is the unknown), then it should become obvious that Boyle's Law should be used (i.e., apply $P_1 V_1 = P_2 V_2$). Make a list—it often helps!

6. In addition to the individual relationships presented earlier, they can be combined together in a single equation. The general/combined gas equation:

$$\frac{P_1V_1}{n_1T_1} = \frac{P_2V_2}{n_2T_2}$$

If the number of moles of gas in an experiment is constant (frequently the case), then $n_1 = n_2$ and, therefore, n_1 and n_2 can be left out of the equation. When this happens, the equation becomes:

$$\frac{P_1V_1}{T_1} = \frac{P_2V_2}{T_2}$$

(In these equations, P_1, V_1, T_1, and n_1 are the original conditions and P_2, V_2, T_2, and n_2 are the new conditions. Units must be the same on each side of the equations, and temperature must be in Kelvin.)

7. Effusion and diffusion are behaviors of gases that are very similar in their actions and their mathematical relationships. Effusion is the process in which a gas escapes from a vessel by passing through a very small opening. Diffusion is the process by which a homogeneous mixture is formed by the random mixing of two different gases.

Test Tip

On the new AP Chemistry exam you are not required to perform calculations on effusion and diffusion, but you should understand qualitatively that larger molecules diffuse more slowly and pass through small openings at lower rates than smaller molecules.

8. Dalton's Law states that the total pressure of a mixture of ideal gases (e.g., gas A, B, and C) is the sum of their partial pressures.

$$P_{\text{Total}} = P_A + P_B + P_C$$

Each gas in a mixture behaves independently from one another in the mixture, and the pressure exerted by each gas can be calculated using the ideal gas equation for each gas.

$P_AV = n_ART$ to calculate the pressure exerted by gas A

$P_BV = n_BRT$ to calculate the pressure exerted by gas B

$P_CV = n_CRT$ to calculate the pressure exerted by gas C

Since V, R, and T for a mixture of gases in a single vessel are the same for all of the gases, then the total pressure in the mixture is calculated using the sum of the pressures, so alternatively,

$$P_{\text{Total}}V = n_{\text{Total}}RT$$

Mole fraction (X)—the number of moles of a particular substance divided by the total number of moles. The mole fraction of a gas, A, in a mixture of gases A, B and C, is given by,

$$X_A = \frac{n_A}{n_A + n_B + n_C} = \frac{n_A}{n_{Total}}$$

If you know the total pressure of the mixture of gases, one can multiply it by the mole fraction to calculate the partial pressure of the gas: $P_A = X_AP_{Total}$.

A common application of Dalton's Law is seen when collecting a gas over water. In the apparatus shown below, a gas evolved from the chemical reaction in the flask, displaces water, and is collected at the top of the gas jar. As the water is pushed out, some of the water forms a vapor itself, and as a result, a small, additional pressure is exerted in the gas jar; i.e., the collected gas contains both the gas of interest *and* a small amount of water vapor. The total pressure in the gas jar is made up of both the gas that forms from the chemical reaction *and* a small amount of water vapor. In order to find the pressure of the gas from the chemical reaction alone, we must subtract the small water vapor pressure from the total pressure.

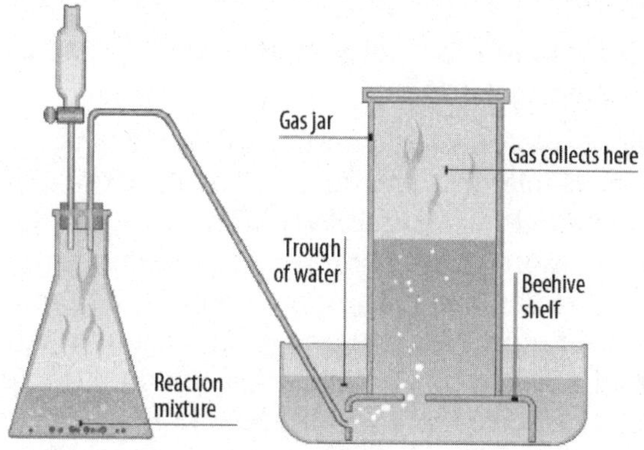

E. Deviations from Ideal Behavior (Real Gases)

1. Gases exhibit ideal behavior as long as pressures are relatively low and temperatures are relatively high.
2. When gases are put under pressure and cooled (i.e., when they move toward condensation into liquids), deviations from ideal behavior are observed and they are said to become *real gases*.
3. The assumptions of KMT regarding the gas particles occupying a negligible volume compared to the whole, and there being no attractions between the particles, both begin to fail, and the ideal gas behaves more like a *real gas*.

4. The larger the gas particles and the stronger the intermolecular interactions between the gas particles, the greater the deviation from ideal behavior.

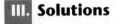

III. Solutions

A. Definitions

1. A solution is a homogeneous mixture—the properties are the same no matter what part of the sample one examines.
2. A solution has a solute and a solvent component, and solutions can be solids, liquids, or gases. For example, a liquid solution can have a gas, liquid, or solid as its solute.
3. Liquid solutions do not scatter visible light and cannot be separated by filtration.
4. Colloids typically have particles that are larger in size than the molecules in a solution, up to approximately 1000 nm. If the particle size exceeds 1000 nm, the mixture becomes a suspension. These particle size distinctions mean that a mixture can appear homogeneous or heterogeneous depending on the scale on which it is observed. For example, with the naked eye, a colloid (like an emulsion paint with a solid dispersed in a liquid) looks homogeneous, but on a hugely magnified scale, it may appear heterogeneous.

B. Separation Techniques

1. Separation techniques depend upon the intermolecular attractions in the solution.

   i. Distillation is the process of heating a liquid mixture and relying upon the difference in boiling points of the components to collect one (the lower boiling point component) above the other, such as distillation of ethanol (boiling point 78°C) and water (boiling point 100°C).

   The technique relies upon the fact that the intermolecular forces between ethanol molecules are weaker than the intermolecular forces between water molecules.

ii. Chromatography is the process of separation of components based upon their relative affinities (intermolecular attraction) for a stationary phase (often paper) and a moving phase (a solvent). Components with a greater affinity for the solvent travel further on the chromatogram.

R_f values (retention factors) are calculated by applying the following formula:

$$R_f = \frac{\text{distance traveled by component of mixture}}{\text{distance traveled by solvent front}}$$

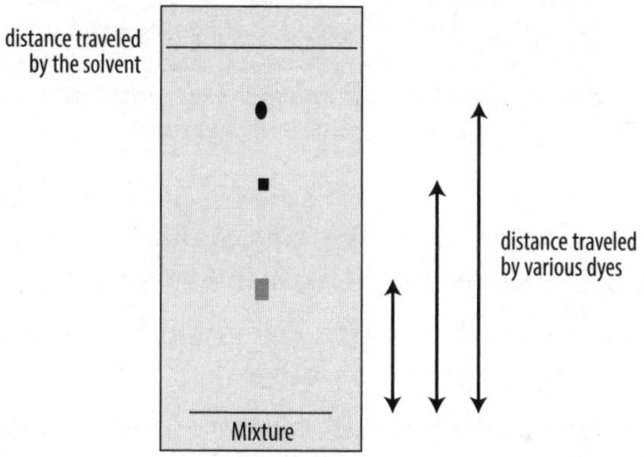

The further a component travels on the chromatogram, the greater its affinity for the mobile phase and the smaller its affinity for the stationary phase. These relative attractions depend upon intermolecular attractions.

C. Preparation of Solutions

1. Solution composition is usually expressed in terms of molarity (M), i.e., the number of moles of solute per liter of solution, and has the units mol L^{-1} or M.
2. Solutions are often prepared by adding water to (diluting) more concentrated ones.

3. In the lab, accurately graduated glassware (volumetric flasks, graduated glass pipets, and burets) are required to prepare such solutions. In general, some water is added to a volumetric flask of appropriate size (i.e., the final volume of solution required). Then, using a suitably sized pipet or a buret, the concentrated solution is slowly and carefully transferred to the volumetric flask with gentle swirling to ensure good mixing. When all of the concentrated solution has been added to the flask, the solution is filled up to the mark of the volumetric flask with solvent.

Test Tip

When diluting strong acids or bases, always add the concentrated acid or base to a large volume of water. This helps dissipate the energy that can be generated by these exothermic solvation processes, thus making for a safer dilution. Adding a small volume of water to a concentrated acid or base is potentially dangerous.

4. Energy changes in solution preparation. Three separate processes take place when a solution is made.
   i. Solute particles must separate from one another (bond breaking process, endothermic).
   ii. Solvent particles must separate from one another (bond breaking process, endothermic).
   iii. Solvent and solute particles must interact (bond making process, exothermic).

   The energy change of the overall process is a sum of energy changes 1, 2, and 3 and can be exothermic or endothermic, depending on the magnitude of each part.

Practice Questions

1. 41.5 L of N_2 gas is collected over water at 23°C. The total pressure of the gases in the collecting flask is 775 mmHg. Calculate the number of moles of N_2 collected. (At 23°C, the vapor pressure of water = 0.028 atm)

2. A mixture of 43.2 g of oxygen gas and 22.1 g of carbon dioxide are contained in a vessel exerting a total pressure of 980 mmHg. Find the pressure exerted by each gas.

3. Calculate the volume of water required to prepare 1.0 L of 2.0 M NaOH from a stock solution that has a concentration of 3.0 mol L^{-1}.

4. Three containers are all filled with equal volumes of three separate gases, O_2, CO_2, and N_2. All of the containers are at the same temperature and pressure.

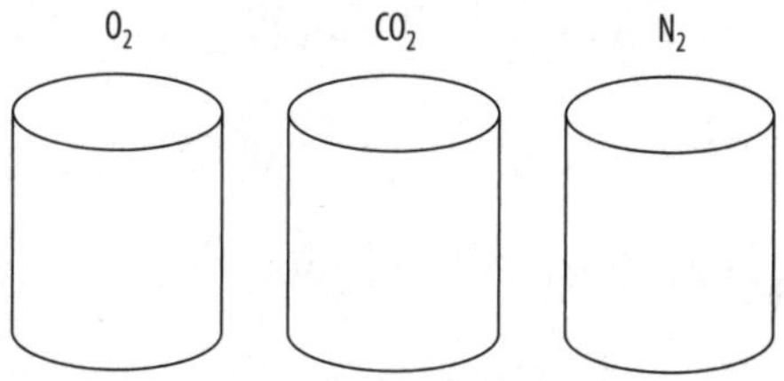

   (a) Which container has the smallest mass?
   (b) In which container do the gas particles have the greatest velocity?
   (c) In which container do the gas particles have the greatest kinetic energy?
   (d) Which gas would leak the fastest out of a hole in the container?

Answers

1. Dalton's Law of Partial Pressure says that

   $775 \text{ mmHg} = 1.02 \text{ atm} = P_{N_2} + P_{H_2O}$

   At 23°C vapor pressure of water = 0.028 atm,

   $1.02 \text{ atm} = P_{N_2} + 0.028 \text{ atm}$

   $P_{N_2} = 0.99 \text{ atm}$

   Use Ideal Gas Law to find total moles:

   $$(0.99 \text{ atm})(41.5 \text{ L}) = (n)\left(0.0821 \frac{L \cdot atm}{K \cdot mol}\right)(296 \text{ K})$$

   $n = 1.69 \text{ mol}$

2. $43.2 \text{ g } O_2 = 1.35 \text{ mol } O_2$

   $22.1 \text{ g } CO_2 = 0.502 \text{ mol } CO_2$

   Total moles = 1.85 mol

   $$P_{O_2} = \frac{1.35 \text{ mol } O_2}{1.85 \text{ mol}}(980 \text{ mmHg}) = 715 \text{ mmHg}$$

   $$P_{CO_2} = 980 \text{ mmHg} - 715 \text{ mmHg} = 265 \text{ mmHg}$$

3. Final solution must contain (1.0 L)(2.0 mol/L) mols = 2.0 mols of NaOH.

   Since moles = (concentration) (volume), the volume (in L) of the stock (concentrated) solution that contains 2.0 mols of NaOH = 0.67 L. So, by taking 0.67 L of the stock solution and adding 0.33 L of water to make the solution up to 1.00 L, the final, diluted solution will have a concentration (molarity) = $\dfrac{2.0 \text{ mol}}{3.0 \text{ mol L}^{-1}}$ = 2.0 mol L^{-1} or 2.0 M.

4. (a) All contain the same number of molecules, but since N_2 has the smallest molar mass, it has the least mass.

   (b) Since N_2 is the lightest, it is traveling the fastest.

   (c) All have the same kinetic energy because they are at the same temperature.

   (d) N_2 would leak the fastest since it has the smallest gas particles (smallest molar mass).

Chapter 8

Intermolecular Forces

I. Intermolecular Forces

A. Induced Dipole–Induced Dipole (London Dispersion Forces)

1. London dispersion forces are small Coulombic (electrostatic) forces that are caused by the movement of electrons within the covalent bonds of molecules that would otherwise have no permanent dipole.
2. As one molecule approaches another, the electrons of one or both are temporarily displaced owing to their mutual repulsion. This movement causes small, temporary, induced dipoles to be set up which attract one another. These attractions are called *London dispersion forces.*
3. These dispersion forces increase with increasing number of electrons in the molecule and with increasing surface area. This leads to greater polarizability, greater attraction, and, therefore, higher melting and boiling points for molecules that contain greater LDFs.
4. London dispersion forces are enhanced in molecules with pi (double and triple) bonds.
5. London dispersion forces exist between *all* molecules.
6. Individually, London dispersion forces are the weakest of the (weak) intermolecular forces, *but* their accumulation in larger molecules (large number of electrons and large surface areas) can lead to them being the strongest *net* force between two molecules. For example, chlorine has as a higher boiling point than HCl, despite chlorine only having relatively weak LDFs compared to HCl's dipole–dipole and LDF interactions. It is the relatively large size of the Cl atoms that increases the

size of the LDFs to a point where they collectively outweigh the combination of dipole–dipole forces and LDFs in HCl.

Test Tip

The College Board has become obsessed with the use of the term "Coulombic attraction" in the new curriculum, which simply means an electrostatic, "opposites attract" force. Don't be confused by the term but expect to see it mentioned a lot on the new exam. Be prepared to answer questions by using the phrase.

B. Dipole–Dipole, Dipole–Induced Dipole, and Hydrogen Bonds

1. Dipole–dipole interactions are Coulombic (electrostatic) attractions between permanent dipoles in adjacent molecules.

   i. Molecules with polar bonds (caused by differences in electronegativity) and dipoles that do not cancel via symmetry will have permanent dipoles.

   ii. When molecules that have permanent dipoles approach one another, they will arrange themselves so that the negative and the positive ends of the molecules attract one another in such a way that maximizes the attraction and minimizes the repulsion.

   iii. Dipole–dipole interactions are intermediate in strength in terms of the (weak) intermolecular forces.

   iv. Polar molecules that have dipole–dipole intermolecular forces tend to have greater attractions than nonpolar molecules of similar size. (Their London dispersion forces are likely to be comparable.)

Dipole–Dipole Attractions

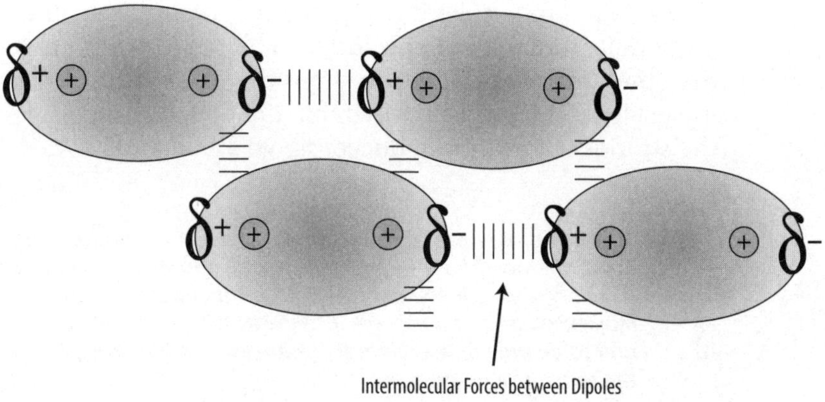

Polar Water Molecules Hydrating Ions and Dissolving an Ionic Salt

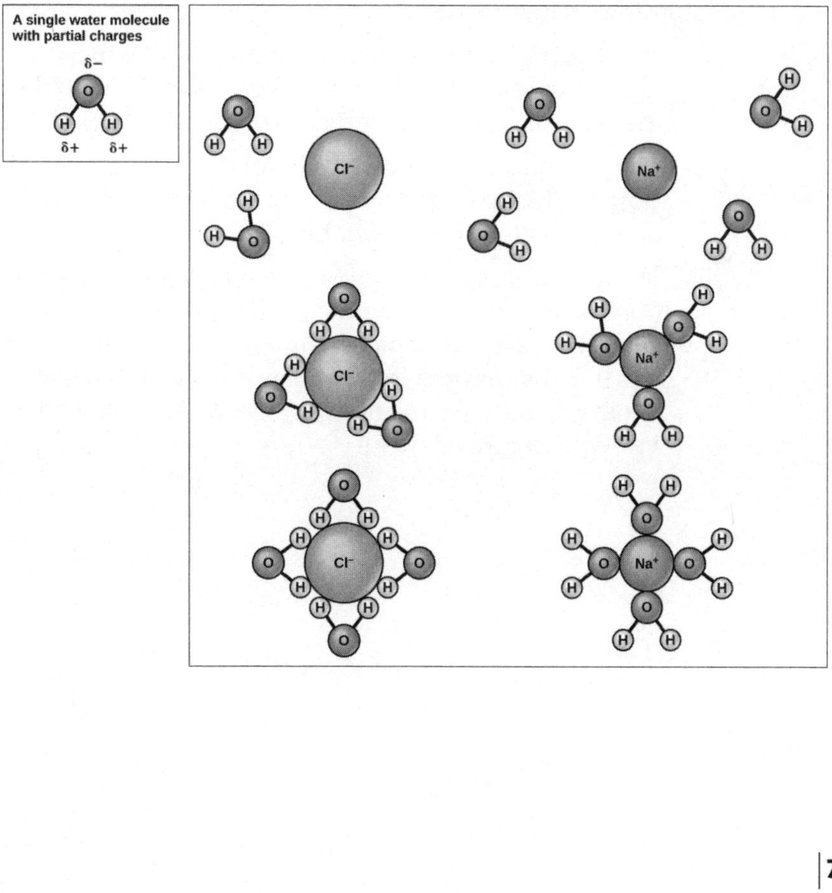

2. Dipole–induced dipole interaction is the same Coulombic (electrostatic) interaction as dipole–dipole, except only one of the pair of molecules has a permanent dipole. The other is nonpolar, but has a dipole induced by the approach of the other polar molecule. The larger the dipole of the polar molecule, and the more polarizable the nonpolar molecule, the stronger the dipole–induced dipole attractions.

Test Tip

Be ready to interpret particulate diagrams of molecules, ions, and other particles interacting with one another. You must be ready and able to identify the individual particles (atoms, molecules, and/or ions), the interparticulate forces at work, and to be able to associate those forces with the properties of the substance in question.

3. Hydrogen bonding is when hydrogen atoms bond to certain elements and create unusually large dipole–dipole, intermolecular attractions.
   i. Hydrogen is an exceptional element in that when it forms a covalent bond, its electron is held to one side of the nucleus, leaving the other side completely exposed.
   ii. Any approaching negatively charged group can get very close to the hydrogen nucleus and this produces an unexpectedly large Coulombic attraction.
   iii. These electrostatic attractions are exaggerated when H is bonded to a more electronegative element that is small enough to allow a significant intermolecular interaction, i.e., fluorine, oxygen, or nitrogen. Such exaggerated, intermolecular, electrostatic attractions are called *hydrogen bonds.*

iv. The occurrence of hydrogen bonds has two important consequences:

- It gives substances containing them anomalously high boiling points.

Hydrogen Halide	Normal Boiling Point, °C
HF*	19
HCl	–85
HBr	–67
HI	–35

*High BP attributed to hydrogen bonding

- Substances containing them tend to have increased viscosity.

Both are explained by the increased attraction between molecules caused by hydrogen bonding, making it more difficult to separate them.

v. Hydrogen bonds can occur between different parts of a single molecule.

vi. Hydrogen bonds are the strongest of the (weak) intermolecular forces.

II. Properties and Intermolecular Forces

A. Boiling Point and Vapor Pressure

1. Strong intermolecular attractions between particles lead to high boiling points and, consequently, low vapor pressures.
2. Assumptions in the Kinetic Molecular Theory that ideal gas particles do not attract one another are incorrect when dealing with real gases, and, as a result, deviations from ideal behavior is observed in gases such as those in group 18. These deviations can be illustrated graphically when pressure versus volume graphs do not adhere to ideal shapes.

B. Surface Tension

1. In the body of a liquid, particles experience forces in three dimensions around them. This results in no net forces on these particles. However, particles at the surface of the liquid have no particles above them, and as such they are pulled with a net force into the body of the liquid. The cohesion between the particles causes the liquid to contract to the smallest possible size (a sphere) and creates an internal pressure at the surface that can resist an external pressure.

C. Capillary Action

1. If a liquid is placed into a very thin tube, the combination of *cohesive* forces within the liquid itself and *adhesive* forces between the liquid and the walls of the tube can add up to overcome the force of gravity, and the liquid can be drawn up the tube.

D. Miscibility and Solubility

1. Two substances with similar intermolecular forces will likely be miscible or soluble in one another (this is sometimes expressed as "like dissolves like").
2. When an ionic salt dissolves in a polar substance such as water, ionic attractions between the ions must be broken, intermolecular forces between the solvent molecules must be broken, and ion to solvent attractions must be made. The combination of these processes and entropic factors determine solubility.

E. Biological Applications

1. The interaction of enzymes with substrates depends upon intermolecular attractions and forces.
2. Proteins, which have areas that are both hydrophobic (water hating) and hydrophilic (water loving), will have three-dimensional shapes that are determined by the attractions (or repulsions) of such areas for water.

3. Base pairs in DNA, A—T (Adenine-Thymine) and G—C (Guanine and Cytosine) are determined through the maximization of hydrogen-bonds at the molecular level. Intermolecular forces are also the reason that DNA is shaped in a double-helix.

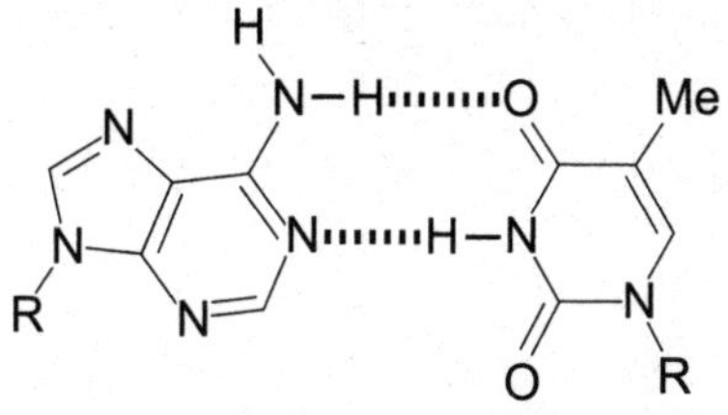

A—T base pair

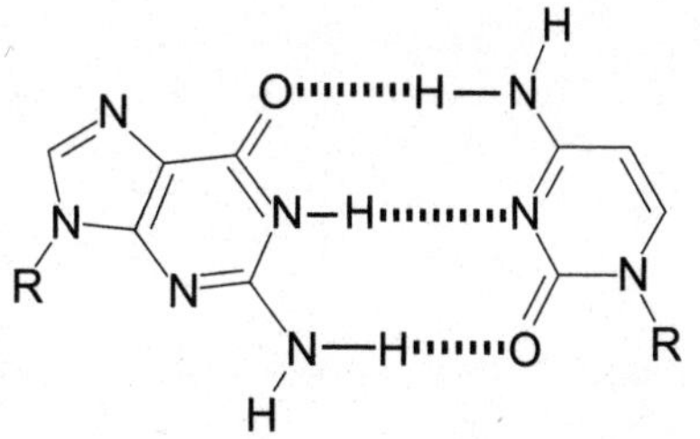

G—C base pair

Intra Forces, VSEPR, and Shape

Chapter 9

I. Chemical Bonds

A. Types and Electronegativity

1. In covalent bonds, atoms share one or more pairs of valence electrons. Covalent bonds are typically formed between two non-metals that are close to one another in the periodic table.
2. In ionic bonding, valence electrons are transferred from one atom to another, and the ionic bond is created due to the strong electrostatic (Coulombic) attraction between the resultant charged particles (i.e., between the ions). Ionic bonds are typically formed between metals and non-metals that are *not* close to one another in the periodic table.
3. Electronegativity is the ability of an atom within a covalent bond to attract electrons to itself. It *increases* left to right on the periodic table and *decreases* down a group. Fluorine is the most electronegative element. The electronegativity of an atom depends on its nuclear charge, the proximity of the electrons that are being attracted to the nuclear charge, and the shielding in that atom.
4. Large differences in electronegativity mean a more uneven sharing of electrons in covalent bonds. Covalent and ionic bonding are two extremes on a continuous spectrum of bonding. Most compounds exhibit some degree of both ionic and covalent characters. By examining properties, it is possible to tell which character dominates.
5. Metallic bonding occurs in metals. Valence electrons are delocalized and are mobile around a regular arrangement of cations.

B. Covalent Bonds

1. Atoms that share electrons are covalently bonded. If the two atoms in a covalent bond have identical electronegativities, then the electrons will be equally shared, and the bond is said to be *nonpolar*. Atoms with very similar (but nevertheless different) electronegativities, like C and H (typically in organic molecules), are often considered to form nonpolar covalent bonds.
2. Nonpolar covalent bonds are formed by minimizing the potential energy between atoms.

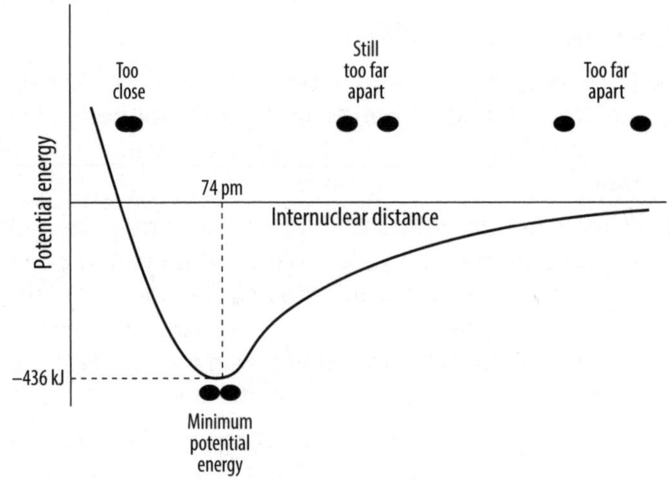

3. Bond length is defined as the distance between the nuclei of the atoms.
4. Bond energy is defined as the energy required to dissociate (break) the bond and is usually reported in units of energy (often kJ) per mole.
5. If the two atoms in a covalent bond have different electronegativities, then the electrons will be unequally shared and the bond is said to be *polar*. The more electronegative atom in the bond will develop a partial negative charge (indicated by $\delta-$), and the less electronegative atom will develop a partial positive charge

(indicated by δ+). The uneven distribution of charge is called a *dipole,* and dipole size increases with increasing differences in electronegativities. An arrow pointing from the partial positive charge (δ+) toward the partial negative charge (δ–) charge is the usual convention used to show the existence of a dipole.

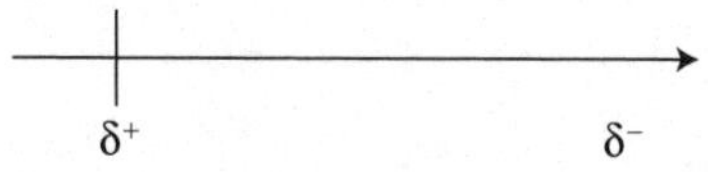

In diatomic molecules, the magnitudes of the two partial charges will be the same.

6. The uneven distribution of these partial charges creates *dipole moments* in molecules. For example, water has a dipole moment illustrated in the following way:

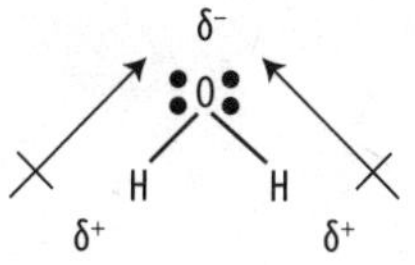

Some molecules have polar bonds but no dipole moment. This is because the individual dipoles cancel out via symmetry and geometry. Two such examples are CO_2 and CCl_4. The canceling of individual dipoles to yield no overall dipole moment is predicted by VSEPR (see VSEPR and shape below).

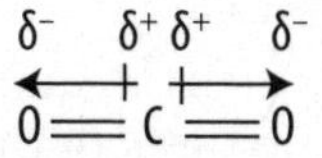

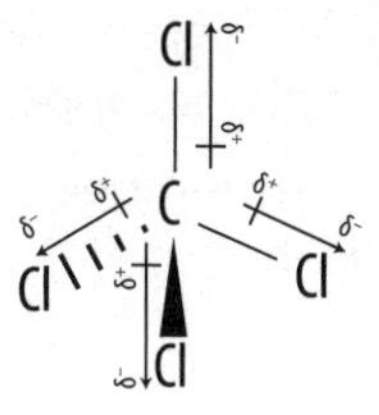

C. Ionic Bonds

1. Cations (positive) and anions (negative) are held together in 3-D lattice structures that maximize the attractions and minimize the repulsions between the ions. The ionic bond is the Coulombic (electrostatic) attraction between the positive and negative ions.

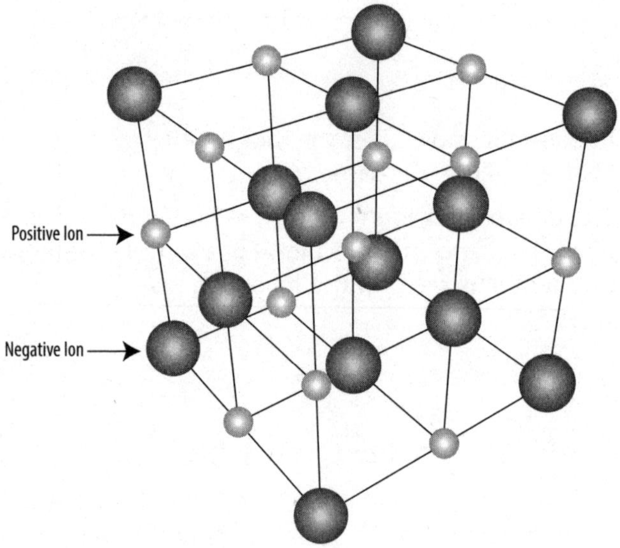

2. Coulomb's law predicts that larger forces and stronger interactions result when ions have greater charges, and when ions can get closer together (i.e., when ions are smaller).

D. Metallic Bonding

1. Metals and alloys (mixtures of metals, see Chapter 10) can be represented as an array of positive ions surrounded by a "sea" of free-moving, delocalized electrons. This structure leads to many characteristic properties such as ductility, malleability, and thermal and electrical conductivity.
2. In general, the greater number of delocalized electrons the stronger the metallic bond.

Typical, Delocalized Metallic Bonding Model

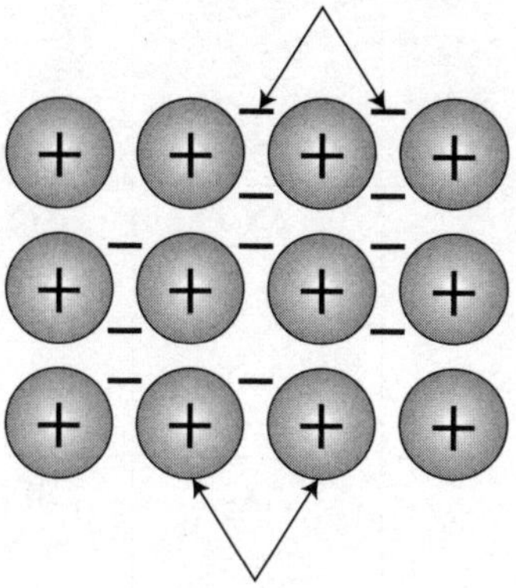

II. VSEPR and Shape

A. VSEPR (Valence Shell Electron Pair Repulsion)

1. VSEPR is a model used to predict the shapes of covalent molecules and polyatomic ions from Lewis diagrams.
2. VSEPR is based upon valence shell electrons repelling each other (via repulsive Coulombic interactions) so that they are as far apart as possible and, therefore, minimize that repulsion.

Molecular Geometry with Central Atom Having No Lone Electron Pair

Molecular Geometry	Number of Lone Electron Pairs Around Central Atom	Formula	Example	VSEPR Prediction and Bond Angle Measurement
Linear	0	AX_2	CO_2	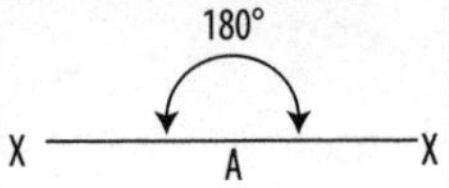
Trigonal Planar	0	AX_3	BF_3	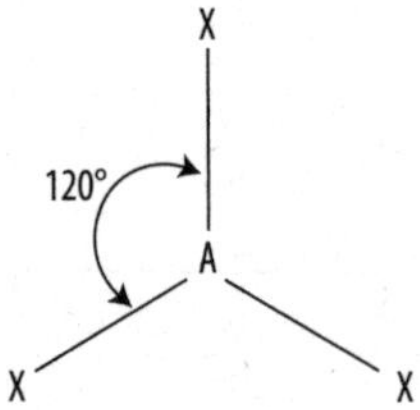
Tetrahedral	0	AX_4	CH_4	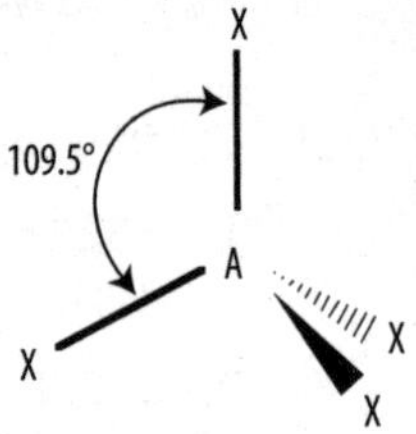
Trigonal-bipyramidal	0	AX_5	PF_5	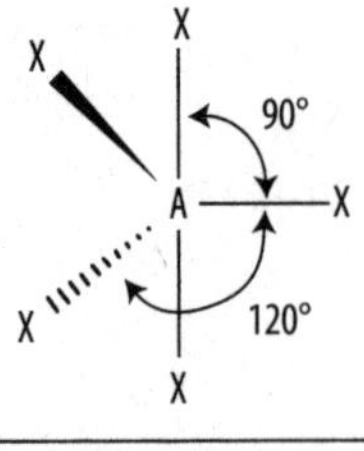
Octahedral	0	AX_6	SF_6	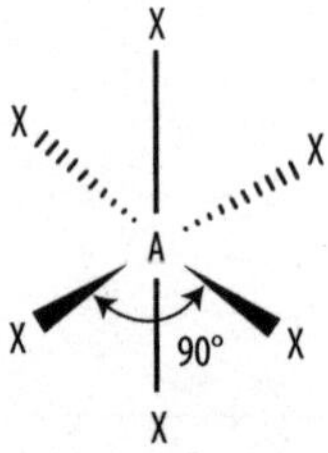

Molecular Geometry with
Central Atom Having One Lone Electron Pair

Molecular Geometry	Number of Lone Electron Pairs Around Central Atom	Formula	Example	VSEPR Prediction
Bent or V-Shaped	1	AX_2E	SeO_2	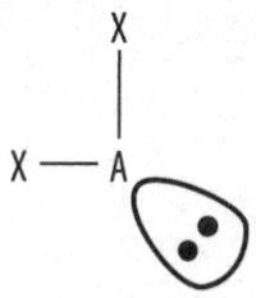
Trigonal Pyramidal	1	AX_3E	NH_3	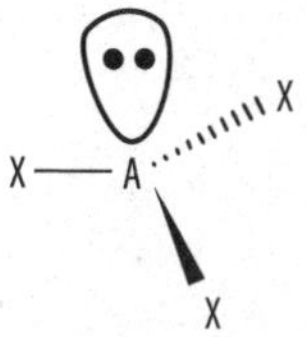
Seesaw	1	AX_4E	SF_4	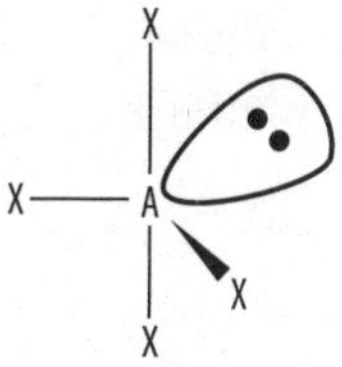
Square-Pyramidal	1	AX_5E	BrF_5	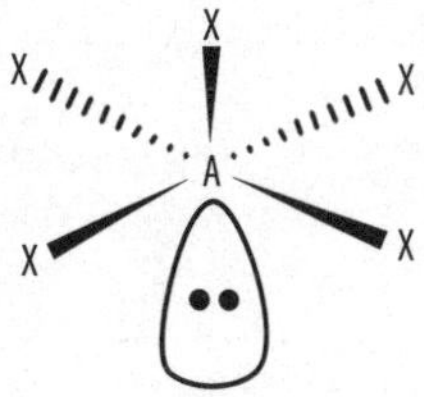

Molecular Geometry with Central Atom Having Two Lone Electron Pairs

Molecular Geometry	Number of Lone Electron Pairs Around Central Atom	Formula	Example	VSEPR Prediction
Bent or V-Shaped	2	AX_2E_2	H_2O	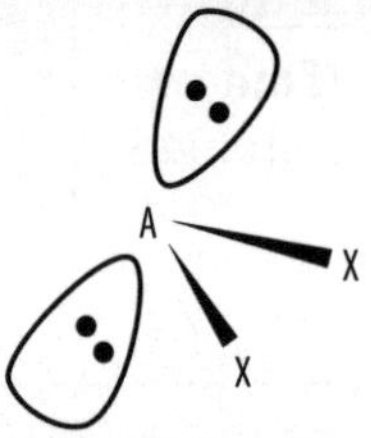
T-shaped	2	AX_3E_2	ICl_3	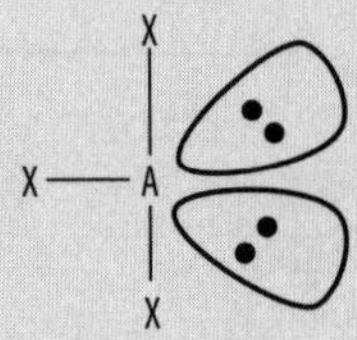
Square Planar	2	AX_4E_2	XeF_4	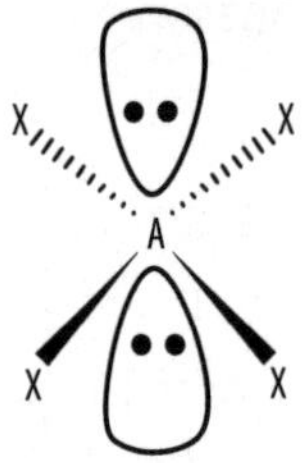

Molecular Geometry with Central Atom Having Three Lone Electron Pairs

Molecular Geometry	Number of Lone Electron Pairs Around Central Atom	Formula	Example	VSEPR Prediction
Linear	3	AX_2E_3	XeF_2	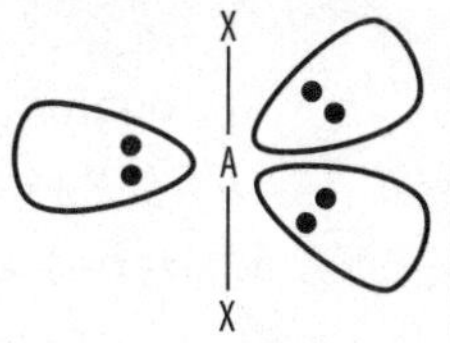

Learning shapes and VSEPR does involve some memorization, but there is also a degree of intuition. For example, three bonding pairs get as far apart as possible (minimize the repulsion) by forming a triangle around the central atom, i.e., molecules with three bonding pairs are trigonal planar.

B. Valence Bond Theory (VBT)
   1. Valence bond theory is based on the concept that bonds are formed by the overlap of atomic orbitals.
      i. Orbitals overlap to form bonds between atoms.
      ii. One electron from each of the bonded atoms accommodates the overlapping orbital.
      iii. Both electrons in the overlap are attracted to the nucleus of each atom. This is why electron pairs are located between two atoms in Lewis structures.
   2. Sigma (σ) bonds (single bonds) are formed when an overlap occurs along the internuclear axis of the bond.
      i. The overlap of two *s* orbitals such as in H-H.
      ii. The overlap of an *s* and *p* orbital such as in H-Cl.
      iii. The overlap of two *p* orbitals such as in Cl-Cl.

3. Pi (π) bonds occur in double and triple bonds. Sideways overlap occurs above and below the internuclear axis of the bond.
   i. The overlap of two *p* orbitals such as in C=C.
   ii. Double bonds contain 1 sigma and 1 pi bond.
   iii. Triple bonds contain 1 sigma and 2 pi bonds.
4. Sigma bonds are generally stronger than pi bonds, so they have larger bond energies. However, multiple (double and triple) bonds contain a combination of sigma and pi bonds, meaning that they tend to be shorter *and* stronger than sigma (single) bonds alone.
5. Hybrid orbitals
   i. Formed by the blending (mixing) of s and p orbitals.
   ii. Hybridization can be matched to molecular geometry/ total number of electrons around the central atom.

Geometry	Number of Negative Centers Around the Central Atom (double and triple bonds count as only one negative center)	Hybridization
Linear	2	sp
Trigonal Planar	3	sp^2
Tetrahedral	4	sp^3

Test Tip

Hybridization beyond sp^3 is no longer tested on the AP Chemistry exam.

6. Some molecules require a more complicated model of bonding than the VBT called the molecular orbital (MO) theory.

C. Resonance and Formal Charge

1. Resonance is used when more than one feasible Lewis structure exists.
2. Resonance often occurs when a double bond can be positioned in more than one place.
3. Bond lengths and strengths found in these species are "averages" of the single and double bonds present.
4. An example is the carbonate ion, as shown in the following diagram. Bond lengths and strengths are found to be intermediate between single and double bonds and have orders of 1⅓ (i.e., four bonds divided amongst three oxygen atoms).

Resonance Structures of the Carbonate Ion CO_3^{2-}

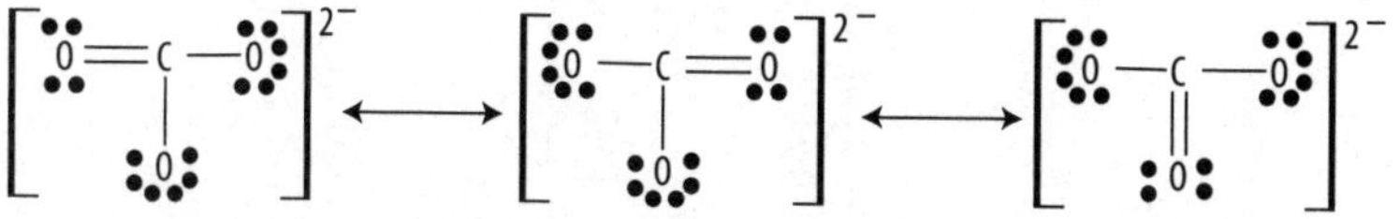

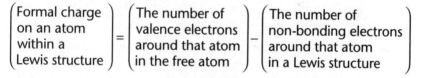

5. Formal charges of each atom within a structure can be calculated using the formula below.

$$\begin{pmatrix}\text{Formal charge}\\ \text{on an atom}\\ \text{within a}\\ \text{Lewis structure}\end{pmatrix} = \begin{pmatrix}\text{The number of}\\ \text{valence electrons}\\ \text{around that atom}\\ \text{in the free atom}\end{pmatrix} - \begin{pmatrix}\text{The number of}\\ \text{non-bonding electrons}\\ \text{around that atom}\\ \text{in a Lewis structure}\end{pmatrix}$$

$$-\frac{1}{2}\begin{pmatrix}\text{The number of}\\ \text{bonding electrons}\\ \text{around that atom}\\ \text{in a Lewis structure}\end{pmatrix}$$

6. The formal charge of an atom within a Lewis structure is generally used in one of two ways:
   i. To suggest where charges may most reasonably lay (for example where the 2– charge of the carbonate ion actually resides)

Using the Lewis structure for the carbonate ion (CO_3^{2-}) as an example, it is possible to determine the formal charges on the atoms and, therefore, the most likely separation of charge.

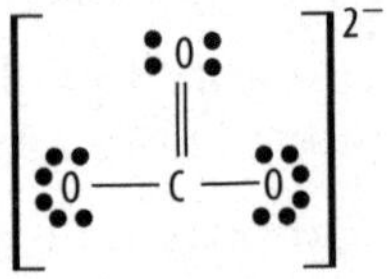

Carbon atom	Formal charge = 4 – 0 – ½ (8) = 0
Oxygen atom in C=O	Formal charge = 6 – 4 – ½ (4) = 0
Each oxygen atom in C-O	Formal charge = 6 – 6 – ½ (2) = –1

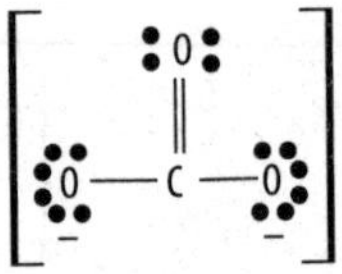

Note that the sum of the formal charges adds up to the total charge on the ion or molecule.

ii. To help select the most plausible structure from a set of resonance structures.

In order to determine which structure is most likely, choose the structure with zero formal charges, and/or formal charges with absolute values as low as possible, and/or keep any negative formal charges on the most electronegative atoms.

Practice Question

1. Based on the Lewis structure below, indicate the hybridization and the number of sigma and pi bonds around the C and N atoms. What shape (molecular geometry) is predicted around each carbon atom?

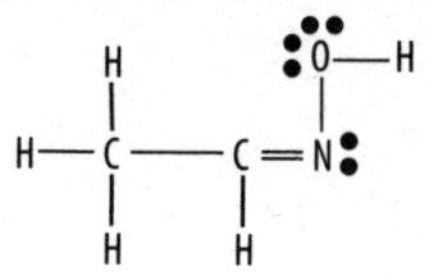

Answer

1. From left to right the first carbon is sp^3 (4 σ bonds), the second carbon is sp^2 (3 σ bonds and 1 π bond), and nitrogen is sp^2 (2 σ bonds and 1 π bond).

   From left to right the first carbon atom is tetrahedral (4 bonding pairs and 0 lone pairs) and the second carbon is trigonal planar (3 bonding pairs—since double bond counts as one—and 0 lone pairs).

Chapter 10

Bonding and Properties of Solids

I. Principles

A. Bonding and Properties

1. The type of interaction (bond) that exists between the particles in any given solid will influence the solid's properties.
2. Properties, such as vapor pressure, conductivity, melting point, hardness and others, can give information about the type of bonds present (and vice versa), and should be viewed as clues that can reveal the type of solid/bonding present.

II. Types of Solids

A. Ionic Solids

1. Atoms have equal numbers of protons and electrons and, consequently, have no overall charge (i.e., they are neutral). When atoms lose or gain electrons in order to achieve a "noble gas structure" and stability) the proton/electron numbers are unbalanced, causing the particles to become charged. These charged particles are called *ions*. Ionic bonds are strong *intra*bonds, caused by the attraction between the oppositely charged ions and are usually formed between metals and non-metals (since metals' atoms have a tendency to lose electrons to form positive ions, and nonmetal atoms have a tendency to gain electrons to form negative ions).
2. The ionic bond (the attractive force between the ions, F) is governed by Coulomb's law and is directly proportional to

the charges on the ions (q_1 and q_2) and inversely proportional to the square of the distance between the ions (r).

$$F \alpha \frac{q_1 q_2}{r^2}$$

i. Ionic solids made up of ions with higher charges (larger q values) will have greater attractions (larger Coulombic forces) and higher melting points.

ii. Smaller ions can get closer together, so they will have smaller values for r and, consequently, greater attractions (larger Coulombic forces) and higher melting points.

iii. It is also possible to define the attractions between two ions in an ionic solid in terms of their *charge density*. The more charge dense the ions, the greater the attraction. This is consistent with i and ii above.

3. Ionic substances form regular, 3-D cubic arrangements called *giant ionic lattices* (as shown below) where the Coulombic forces are strong. As a result of the strong forces, ionic solids tend to have low vapor pressures, high melting points, and high boiling points.

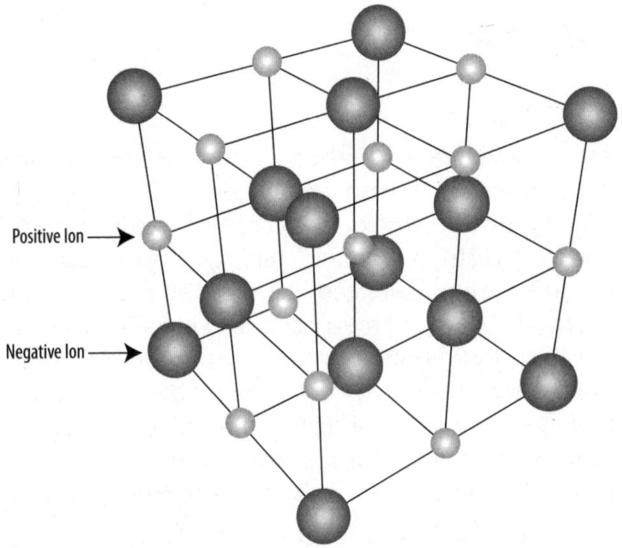

4. Ionic substances are brittle because like charges repel one another when the layers slide over one another.
5. Ionic substances can only conduct electricity when molten or in solution because when they are in the solid state, the ions are rigidly held, can gently vibrate, but cannot move in relation to one another. When molten or in solution, the ions are free to move around one another so the conduction of electricity is possible.
6. The charged particles present in an ionic solid explain how and why ionic solids dissolve in water. When an ionic solid dissolves, the *polar* water molecules penetrate the lattice and attach themselves to the ions. The process is called *hydration* and the ions are said to be *hydrated. Nonpolar* solvents will have no such attraction for the ions, and the strong forces between the ions are not interrupted.

Test Tip

Expect to encounter many particulate diagrams like the one below on the new AP exam.

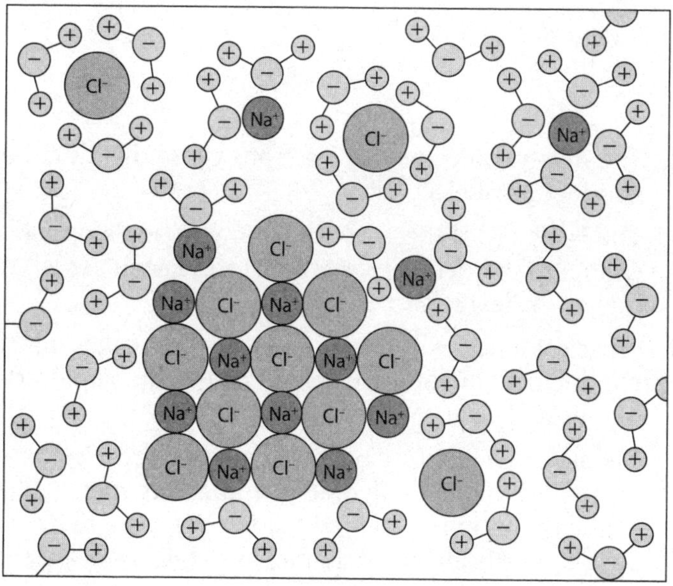

You may have heard the phrase "like dissolves like." This refers to the fact that charged substances (like ionic solids) tend to dissolve in polar (charged) solvents. Remember the phrase—it's a good rule of thumb.

B. Metals

1. A metal's bonding and structure can be considered to be a close-packed lattice of positive kernels, surrounded by a "sea" of free-moving valence electrons, as shown in the following diagram.

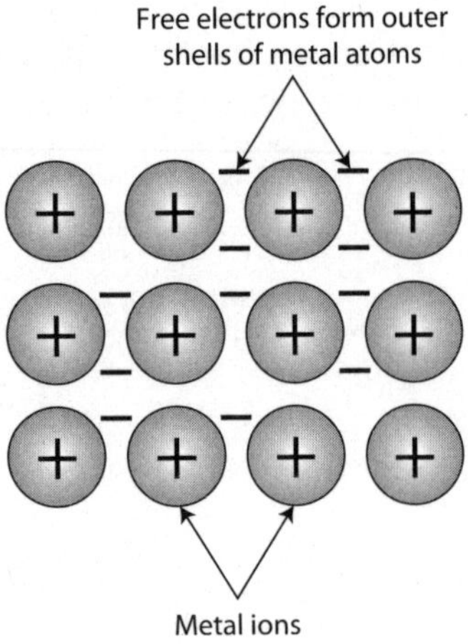

2. The free-moving valence electrons cause metals to be good conductors of electricity.
3. The flexibility of these bonds (they can be deformed without changing the general structure around each kernel) makes metals malleable and ductile.
4. The close-packed nature of the structure makes metals good conductors of heat since energy can easily be passed from one positive kernel to the next.
5. Metals and alloys (see below) may have some surface properties that are different from the pure metal(s), due to the formation of an inert oxide layer, via a reaction with oxygen in the air. For example, in stainless steel, a chromium

oxide layer is formed on the surface of the steel that helps prevent corrosion.

Look out for questions that ask about the properties of metals related to their bonding and structure. Good conduction of heat and electricity, malleability and ductility, along with luster (shininess) are all typical properties of metals. Recognizing these traits can help you answer test questions.

6. Alloys are mixtures of metals.
   i. Interstitial alloys—small atoms fill the spaces (interstices) between the bigger atoms, making the alloy more rigid and less malleable than the pure metal. For example, carbon is added to iron to make the alloy steel.
   ii. Substitutional alloys—atoms of a comparable size to the pure metal, substitute themselves for those pure metal atoms. For example, brass is made from copper and zinc. The alloy will have a density that is somewhere between the densities of the two pure component metals and, like interstitial alloys, they tend to be less malleable than the pure metals.
   iii. Alloys are typically still good conductors, since, like pure metals, they still have a sea of free-moving electrons.

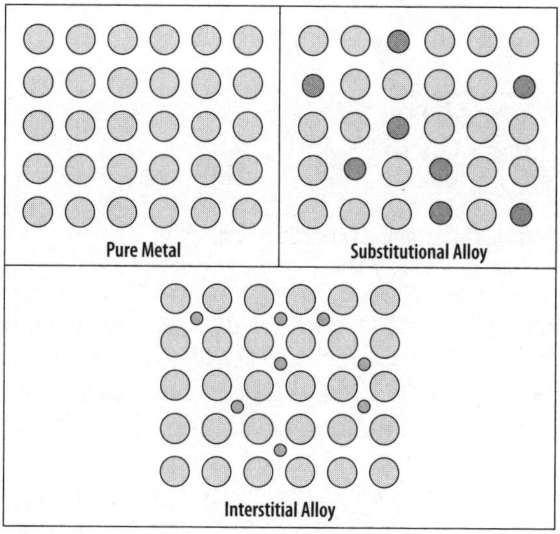

C. Covalent Network Solids

1. These are structures where non-metal atoms are bonded to one another with strong covalent bonds in a massive (or giant), two- or three-dimensional, continuous network. Examples include diamond, graphite, silicon dioxide, and silicon carbide. Group 14 elements are good candidates for such structures since they can form four covalent bonds. Melting points are high and the solids are hard because of the strength of the covalent bonds.
2. The 2-D and 3-D structures are rigid and hard since the bonds are strong and the bond angles are fixed and do not vary or change.
3. Diamond is based upon a 3-D, tetrahedral unit where all of the carbon atoms sp^3 hybridized and are bonded to four other carbon atoms with very strong covalent bonds in a huge macrostructure. Large numbers of covalent bonds make a diamond very strong and hard, and gives it a high melting and boiling point.

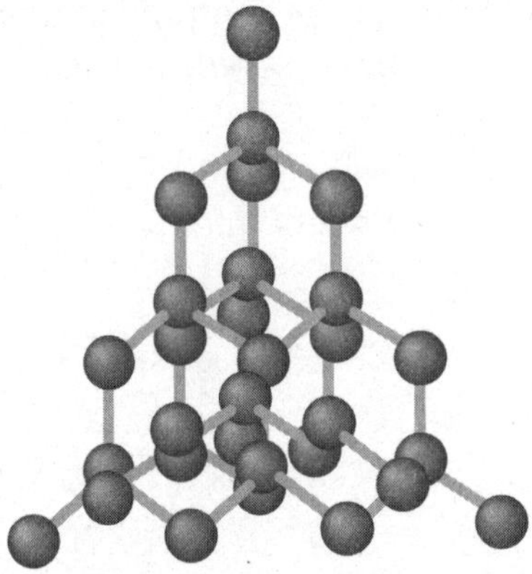

Diamond's Network Covalent Structure

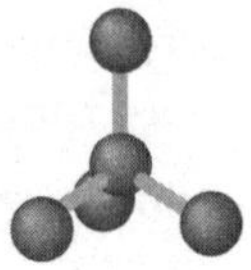

Tetrahedral Unit

4. Graphite has a 2-D, sheet structure, where each carbon atom is sp^2 hybridized and covalently bonded to three other carbon atoms in a plane. The strong covalent bonds give graphite a high melting point. The sheets stack in a layered structure. The layered structure leads to specific properties:

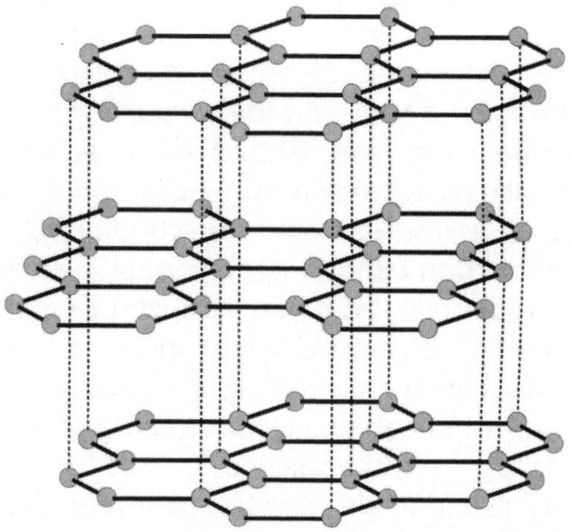

Graphite's Layered Structure

i. It will conduct electricity only in one plane. In the graphite structure, each carbon is only bonded to three others. This leaves one of each carbon atom's valence electrons "free." These electrons are spread out over each layer (delocalized). This leads to a sea of electrons similar to that in metals and is responsible for graphite's ability to conduct electricity along the layers. Because the electrons cannot move between layers, there is no conduction from one layer to another.

ii. It is soft and can be used as a lubricant. Weak London dispersion forces hold the layers in graphite together. As a result, they can slide over one another, making graphite a good lubricant.

5. Silicon forms a similar 3-D structure to diamond.

   i. Silicon is a semiconductor. The doping of silicon by introducing other elements causes the formation of both n-type and p-type semiconductors. N-type (negative) are created by adding atoms of elements from group 15 that have an extra valence electron compared to Si, and p-type (positive) by introducing atoms of elements from group 13 with one less valence electron than Si.

ii. Silicon's conductivity increases with an increase in temperature. In the electronic structure of solids, there are two "bands" of permitted energy levels for electrons: the *valence band* and the *conduction band*. Electrons reside in the valence band, but in order for conduction to take place, they must be promoted to the conduction band. In insulators and semiconductors, the valence and conduction bands are separated by an energy gap, and, in insulators, this gap is large and cannot be bridged. As a result, insulators cannot conduct electricity. In a semiconductor such as silicon, the energy gap is smaller, and, with heating, electrons can gain sufficient energy to cross from the valence band to the conduction band. As a result, at elevated temperatures, silicon's electrical conductivity is increased.

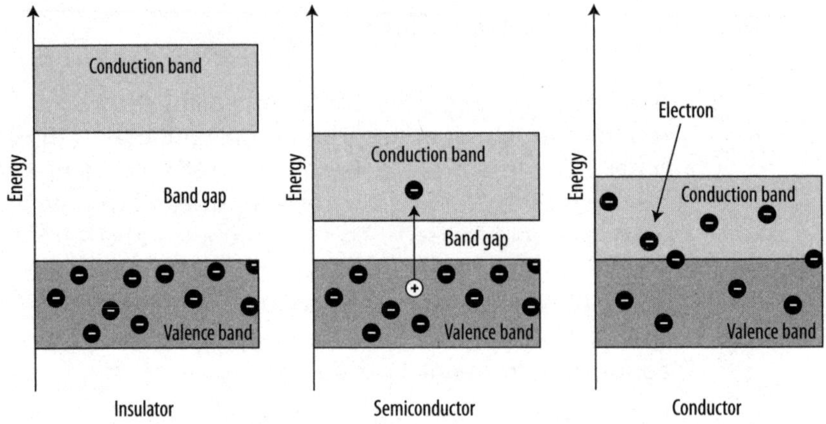

D. Molecular Solids

1. Solids occur when discrete molecules, made from nonmetals, are attracted to one another with only London dispersion forces. An example is iodine.

2. London dispersion forces mean that molecular solids tend to have relatively low melting points because there is only a relatively weak (for solids) interaction between its molecules. Also, they tend to be soft.

3. No ions and no free-moving electrons are present (since electrons are tied up in covalent bonds), so molecular solids are expected to be non-conductors.
4. Some important large molecules and polymers in biology are observed to be molecular solids. Examples include polyethylene and biopolymers, such as polyhydroxybutyrate.

Test Tip

Most test questions expect you to be able to relate observed physical properties, such as melting and boiling points, electrical conductivity, hardness, and water solubility, to the type of structure and bonding present. Focus your review on the relationships between structure and properties.

PART IV

CHEMICAL REACTIONS

Chapter 11

Chemical Reactions

I. Stoichiometry

A. Balancing of Equations

1. Chemical equations must be balanced in order to conserve atoms and, therefore, conserve mass.
2. The relationship of the coefficients between reactants and products in a balanced chemical equation is called stoichiometry.
3. Rules to balance chemical equations:
   i. The same number and type of atom(s) must be present in the reactants and products.
   ii. Coefficients are placed in front of the chemical formula. Do not change the subscripts in (correct) chemical formula.
   iii. *Sometimes* (but not always) it is necessary to have all coefficients expressed as the lowest possible integers.

   The process of balancing equations is not necessarily a "scientific one" (despite the need to have balanced equations being scientific), but the following may help you to move away from a purely "trial and error" process and toward a more structured approach.

   i. First balance metals, then non-metals.
   ii. Balance hydrogen and oxygen last.
   iiii. Balance polyatomic ions as units.
   iv. Be sure to balance charges.

4. What is the anatomy of a chemical equation?

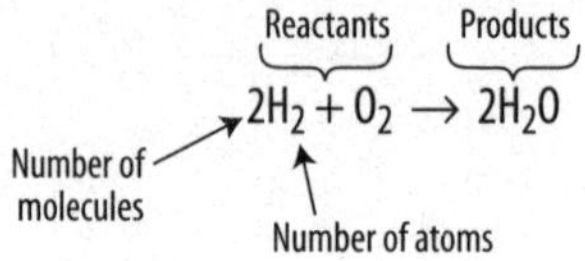

2 molecules of H_2 and 1 molecule of O_2 produce 2 molecules of H_2O

2 mole of H_2 and 1 mole of O_2 produce 2 mole of H_2O

B. **Practice Questions**

Balance the following equations (easy).

1. $\_NaI + \_Pb(SO_4)_2 \rightarrow \_PbI_4 + \_Na_2SO_4$
2. $\_VF_5 + \_HI \rightarrow \_V_2I_{10} + \_HF$
3. $\_Sr(NO_3)_2 + \_GaPO_4 \rightarrow \_Sr_3(PO_4)_2 + \_Ga(NO_3)_3$
4. $\_C_2H_6 + \_O_2 \rightarrow \_CO_2 + \_H_2O$
5. $\_Hg_2I_2 + \_O_2 \rightarrow \_Hg_2O + \_I_2$

Answers

1. 4, 1, 1, 2
2. 2, 10, 1, 10
3. 3, 2, 1, 2
4. 2, 7, 4, 6*
5. 2, 1, 2, 2

(Coefficients of 1 are usually omitted from the final equations, since they are "understood.")

---

*For question #4, it may be easier to balance with fractions first and then multiply by 2 to get the lowest whole-number coefficients. Balancing using fractions is a useful skill, and can make the balancing process easier, but be sure to check that the question does *not* require using the smallest possible integers before presenting a final answer with fractional coefficients. For example, the following equations are both "balanced," but one includes the use of fractions as coefficients, while the other uses the smallest possible integers. Some questions may require the use of only the smallest possible integers.

$\underline{1}C_2H_6 + \underline{3.5}O_2 \rightarrow \underline{2}CO_2 + \underline{3}H_2O$ OR $\underline{2}C_2H_6 + \underline{7}O_2 \rightarrow \underline{4}CO_2 + \underline{6}H_2O$

Balance the following equations (more difficult).

6. $\_NH_3 + \_Br_2 \rightarrow \_N_2 + \_NH_4^+ + \_Br^-$
7. $\_SCl_2 + \_NH_3 \rightarrow \_S_4N_4 + \_NH_4Cl + \_S$
8. $\_NO + \_I^- + \_H^+ \rightarrow \_NH_4^+ + \_I_2 + \_H_2O$
9. $\_FeCr_2O_4 + \_Na_2CO_3 + \_O_2 \rightarrow \_Na_2CrO_4 + \_Fe_2O_3 + \_CO_2$
10. $\_Cu + \_HNO_3 \rightarrow \_NO + \_Cu(NO_3)_2 + \_H_2O$

Answers

6. 8, 3, 1, 6, 6
7. 6, 16, 1, 12, 2
8. 2, 10, 12, 2, 5, 2
9. 4, 8, 7, 8, 2, 8
10. 3, 8, 2, 3, 4

Stoichiometry Flow Chart

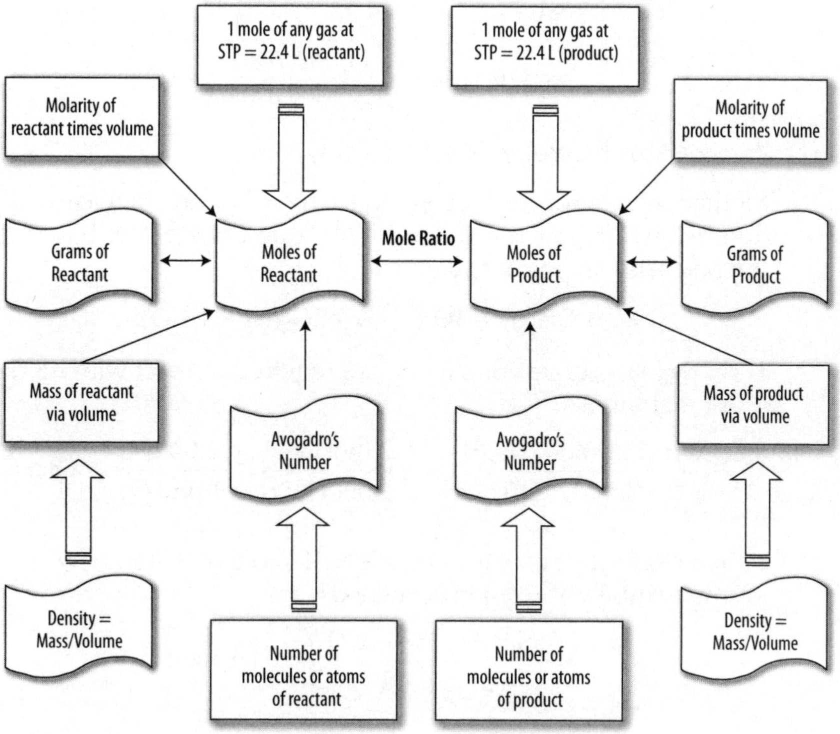

C. The Mole Concept, Including Limiting Reactants and % Yield

1. Mole—the same number of particles as there are atoms, in 12 grams of pure carbon –12. It is represented by Avogadro's number, 6.022×10^{23}.
2. Calculate moles by dividing mass in g by molar mass in g mol^{-1}.
3. Limiting reagent/reactant is the reactant that will be completely used up during the chemical reaction and that determines all other amounts.
4. Excess reagent/reactant is the reactant that will not be completely used up during the chemical reaction and is left over at the end of the reaction.
5. Percent yield is a method to calculate the effectiveness of a chemical reaction.
   i. Actual yield is what you produce from actually doing the reaction in a laboratory setting.
   ii. Theoretical yield is what "should have been" produced from the chemical reaction. This is calculated via stoichiometry.

$$\text{Percent Yield} = \frac{\text{Actual Yield}}{\text{Theoretical Yield}} \times 100$$

D. Practice Stoichiometry Questions (easy)

Methanol is combusted according to the balanced chemical equation below. Answer the stoichiometry questions below based on the balanced chemical reaction.

$$2CH_3OH(g) + 3O_2(g) \rightarrow 2CO_2(g) + 4H_2O(g)$$

1. How many grams of oxygen are required to react with 45 g of methanol?

$$\frac{45\text{ g }CH_3OH}{1} \times \frac{1\text{ mol }CH_3OH}{32.042\text{ g }CH_3OH} \times \frac{3\text{ mol }O_2}{2\text{ mol }CH_3OH} \times \frac{32.00\text{ g }O_2}{1\text{ mol }O_2} = 68\text{ g }O_2$$

2. If a chemist desires to produce 0.95 mole of water, how many moles of oxygen is needed?

$$\frac{0.95\text{ mol }H_2O}{1} \times \frac{3\text{ mol }O_2}{4\text{ mol }H_2O} = 0.71\text{mole }O_2$$

3. If you have 34.5 g of methanol and 62.2 g of oxygen, how much water will be produced?

   Do the calculation twice. First assume methanol to be limiting, and then assume oxygen to be limiting.

$$\frac{34.5\text{ g }CH_3OH}{1}\times\frac{1\text{ mol }CH_3OH}{32.042\text{ g }CH_3OH}\times\frac{4\text{ mol }H_2O}{2\text{ mol }CH_3OH}\times\frac{18.016\text{ g }H_2O}{1\text{ mol }H_2O}=38.8\text{ g }H_2O$$

$$\frac{62.2\text{ g }O_2}{1}\times\frac{1\text{ mol }O_2}{32.00\text{ g }O_2}\times\frac{4\text{ mol }H_2O}{3\text{ mol }O_2}\times\frac{18.016\text{ g }H_2O}{1\text{ mol }H_2O}=46.7\text{ g }H_2O$$

   38.8 g H_2O will be produced.

   i. Determine the limiting and excess reagent.

      a. Limiting reagent is CH_3OH since it produces the smaller amount in the calculations above.

      b. Excess reagent is O_2. If methanol is limiting, then, by definition, oxygen is in excess.

   ii. How much of the excess reagent remains unreacted?

      a. Since 38.8 g of H_2O is produced, work backwards to find how much O_2 was used.

$$\frac{38.8\text{ g }H_2O}{1}\times\frac{1\text{ mol }H_2O}{18.016\text{ g }H_2O}\times\frac{3\text{ mol }O_2}{4\text{ mol }H_2O}\times\frac{32.00\text{ g }O_2}{1\text{ mol }O_2}=51.7\text{ g }O_2$$

      51.7 g O_2 was used.

      62.2 g O_2 – 51.7g O_2 = 10.5 g O_2 UNUSED or IN EXCESS

   iii. When the combustion of methanol was repeated by a chemistry student, 12.4 g of water were produced. What is the percent yield?

$$\text{Percent Yield}=\frac{12.4\text{ g }H_2O}{38.8\text{ g }H_2O}\times 100=32.0\%$$

E. Practice Stoichiometry Questions (difficult)

   Household bleach can be produced based on the chemical reaction below.

   2 NaOH (aq) + Cl_2 (g) → NaOCl (aq) + NaCl (aq) + H_2O (l)

1. What volume of 0.25 M NaOH is required to produce 18 g of NaOCl?

$$\frac{18\text{ g NaOCl}}{1}\times\frac{1\text{ mol NaOCl}}{74.44\text{ g NaOCl}}\times\frac{2\text{ mol NaOH}}{1\text{ mol NaOCl}}\times\frac{1\text{ L NaOH}}{0.25\text{ mol NaOH}}=1.9\text{ L NaOH}$$

2. What volume of 0.25 M NaOH is required to react with 0.32 moles of chlorine?

$$\frac{0.32\text{ mol Cl}_2}{1}\times\frac{2\text{ mol NaOH}}{1\text{ mol Cl}_2}\times\frac{1\text{ L NaOH}}{0.25\text{ mol NaOH}}=2.6\text{ L NaOH}$$

3. If 4.00 L of 0.25 M NaOH are used with excess chlorine gas, how much NaOCl is produced?

$$\frac{4.00\text{ L NaOH}}{1}\times\frac{0.250\text{ mol NaOH}}{1\text{ L NaOH}}\times\frac{1\text{ mol NaOCl}}{2\text{ mol NaOH}}\times\frac{74.44\text{ g NaOCl}}{1\text{ mol NaOCl}}=37.2\text{ g NaOCl}$$

F. Sample Calculations Using Density

1. Liquid bromine is reacted with aluminum to produce aluminum bromide.

$$2\text{ Al(s)} + 3\text{ Br}_2\text{(l)} \rightarrow 2\text{AlBr}_3\text{(s)}$$

If 450. ml of bromine reacts with excess aluminum, how much aluminum bromide will be made? The density of liquid bromine is 3.1 g/mL.

$$\frac{450\text{ mL Br}_2}{1}\times\frac{3.10\text{ g Br}_2}{1\text{ mL Br}_2}\times\frac{1\text{ mol Br}_2}{159.8\text{ g Br}_2}\times\frac{2\text{ mol AlBr}_3}{3\text{ mol Br}_2}\times\frac{266.68\text{ g AlBr}_3}{1\text{ mol AlBr}_3}=1550\text{ g AlBr}_3$$

2. The density of an unknown gas at STP is 0.870 g/L. Find the molar mass of the gas.

$$0.870\frac{\text{g}}{\text{L}}=\frac{(1\text{ atm})(\text{MM})}{\left(0.0821\frac{\text{L}\cdot\text{atm}}{\text{K}\cdot\text{mol}}\right)(273\text{ K})}$$

$$\text{MM}=19.5\frac{\text{g}}{\text{mol}}$$

3. At 1.0 atmosphere and 0°C, calculate the density of NO_2 gas (MM = 46.01 g/mol). (The key here is STP, where 1 mole of any gas at STP = 22.4 L.)

$$\frac{46.01\ \text{g/mol}}{22.4\ \text{L/mol}} = 2.05\frac{\text{g}}{\text{L}}$$

G. Gas Stoichiometry

1. Flowchart

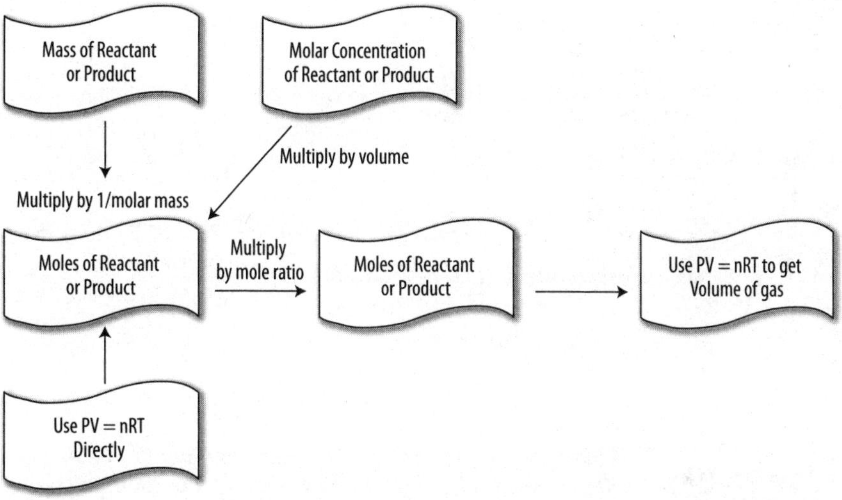

2. Sample stoichiometry and volume problem

An excess of Mg(s) is added to 100 mL of 0.300 M HF. At 0°C and 1 atm pressure, what is the volume of hydrogen gas produced?

Write the balanced chemical equation first.

$$Mg(s) + 2HF\ (aq) \rightarrow MgF_2(aq) + H_2\ (g)$$

Use stoichiometry.

$$\frac{0.100\ \text{L HF}}{1} \times \frac{0.300\ \text{mol HF}}{1\ \text{L HF}} \times \frac{1\ \text{mol}\ H_2}{2\ \text{mol HF}} = 0.0150\ \text{mol}\ H_2$$

1 mol of any gas at STP occupies a volume of 22.4 L.

$$\frac{0.0150\ \text{mol}\ H_2}{1} \times \frac{22.4\ \text{L}\ H_2}{1\ \text{mol}\ H_2} = 0.336\ \text{L}\ H_2$$

3. Sample moles and molar mass problem

   0.20 g of a gas that occupies a volume of 1.0 L exerts a pressure of 1.2 atm at 27.4°C. Find the molar mass of the compound.

   Use PV = *n*RT, the definition of molar mass as being the mass in grams per mole of substance, and remember to convert temperature to K in gas calculations (see Chapter 7).

$$(1.2\ \text{atm})(1.0\ L) = (n)\left(0.0821\frac{\text{L}\cdot\text{atm}}{\text{K}\cdot\text{mol}}\right)(300.4\ K)$$

$$n = 0.049\ \text{mol}$$

$$\frac{0.20\ \text{g}}{0.049\ \text{mol}} = 4.1\frac{\text{g}}{\text{mol}}$$

Test Tip

It is possible to rearrange the equation PV = nRT into other useful forms, such as Molar Mass = $\frac{(mass)(R)(T)}{(P)(V)}$.

Test Tip

Stoichiometry questions can be asked in a multitude of different ways, depending on how the AP question writers want to challenge you. If you use dimensional analysis and your factor label techniques, stoichiometry can actually be a relatively easy part of the AP exam.

II. Molecular, Ionic, and Net Ionic Equations

A. Precipitation Reactions

1. Different circumstances/reactions call for equations to be written in different forms. One such situation is the double displacement, precipitation reaction.
2. Reactants are generally water-soluble ionic compounds that will dissociate into anions and cations.

3. The combination of one of the aqueous anions and one of the aqueous cations will produce the insoluble precipitate.
   i. An example of a precipitation reaction is the one between aqueous solutions of silver nitrate and potassium chloride to give a precipitate of silver chloride.
   ii. The full equation (the equation that shows all of the full formula) is:

$$AgNO_3(aq) + KCl(aq) \rightarrow AgCl(s) + KNO_3(aq)$$

Test Tip

You only need to know a few solubility rules for the AP Chemistry exam. For past exams, you had to know them all, but it is now sufficient to simply know that sodium, potassium, ammonium, and nitrate salts are soluble. If you need other solubility rules, they should be provided in the question.

B. Ionic Equations and Net Ionic Equations
   1. Ionic equation—Using the example above and starting with the full equation, split any aqueous substances into their ions, but leave any non-aqueous substances intact.

$$Ag^+(aq) + NO_3^-(aq) + K^+(aq) + Cl^-(aq) \rightarrow AgCl(s) + K^+(aq) + NO_3^-(aq)$$

   2. Spectator ions are ions that are found in exactly the same form on both sides of the equation, i.e., they do not change/take part in the reaction so they are "spectators." They must be removed when writing the net ionic equation.
   3. Net ionic equation is a balanced chemical equation that is written leaving out the spectator ions.

$$Ag^+(aq) + Cl^-(aq) \rightarrow AgCl(s)$$

C. Example of Writing a Net Ionic Equation

1. The reaction of aqueous silver nitrate with aqueous calcium chloride gives a precipitate of silver chloride.

    i. Balance the full equation.

    $$2AgNO_3(aq) + CaCl_2(aq) \rightarrow 2AgCl(s) + Ca(NO_3)_2(aq)$$

    ii. Write the ionic equation.

    $$2Ag^+(aq) + 2NO_3^-(aq) + Ca^{2+}(aq) + 2Cl^-(aq) \rightarrow$$
    $$2AgCl(s) + Ca^{2+}(aq) + 2NO_3^-(aq)$$

    iii. Locate the spectator ions and remove them to write the net ionic equation.

    $$2Ag^+(aq) + 2Cl^-(aq) \rightarrow 2AgCl(s)$$

    or

    $$Ag^+(aq) + Cl^-(aq) \rightarrow AgCl(s)$$

Test Tip

It is vital to balance the charge as well as the atoms. This is a recurring theme on the AP exam, especially when dealing with REDOX equations.

Practice Questions

Write the ionic and net ionic equations for the following balanced, full equations.

1. $2KI(aq) + Pb(NO_3)_2(aq) \rightarrow 2KNO_3(aq) + PbI_2(s)$

2. $Ba(NO_3)_2(aq) + Na_2C_2O_4(aq) \rightarrow 2NaNO_3(aq) + BaC_2O_4(s)$

Answers

1. Ionic: $2K^+(aq) + 2I^-(aq) + Pb^{2+}(aq) + 2NO_3^-(aq) \rightarrow 2K^+(aq) + 2NO_3^-(aq) + PbI_2(s)$

   Net Ionic: $Pb^{2+}(aq) + 2I^-(aq) \rightarrow PbI_2(s)$ (spectator ions K^+ and NO_3^- removed)

2. Ionic: $Ba^{2+}(aq) + 2NO_3^-(aq) + 2Na^+(aq) + C_2O_4^{2-}(aq) \rightarrow 2Na^+(aq) + 2NO_3^-(aq) + BaC_2O_4(s)$

   Net Ionic: $Ba^{2+}(aq) + C_2O_4^{2-}(aq) \rightarrow BaC_2O_4(s)$ (spectator ions Na^+ and NO_3^- removed)

III. Particulate Representations

A. "Equations" in Diagrammatic Form

1. The reaction between solid carbon C and oxygen gas O_2 to produce carbon monoxide gas CO may be represented as a "balanced equation" in diagrammatic form as follows:

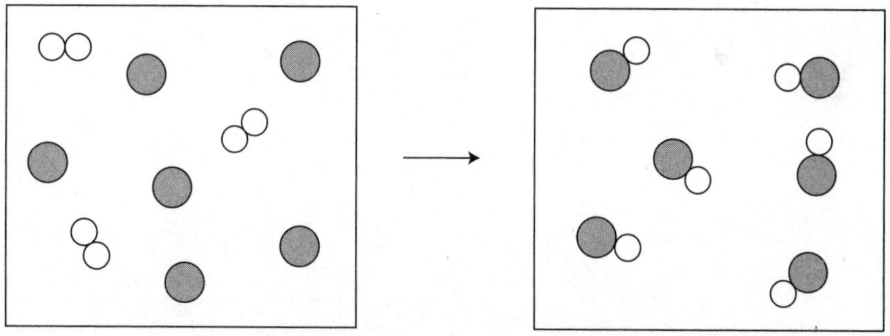

$$6C(s) + 3O_2(g) \rightarrow 6CO(g)$$

Chapter 12

Classifying Chemical Reactions

I. Types of Chemical Reactions

A. Synthesis Reactions

1. Atoms and/or molecules combine to form a new compound.
2. A + B → AB
    i. Metal + non-metal → binary ionic compound

    e.g., $Mg(s) + Cl_2(g) \rightarrow MgCl_2(s)$

    ii. Non-metal + non-metal → binary covalent compound

    e.g., $H_2(g) + F_2(g) \rightarrow 2HF(g)$

    iii. Compound + compound → compound

    e.g., $CO_2(g) + CaO(s) \rightarrow CaCO_3(s)$

B. Decomposition Reactions

1. A compound breaks down into two or more substances.
2. AB → A + B (the reverse of synthesis)
3. Thermal decomposition (using heat to cause decomposition)
    i. Oxy-acids, when heated (and sometimes without the need for heat), decompose to form water and non-metal oxide, e.g., $H_2CO_3(aq) \rightarrow H_2O(l) + CO_2(g)$

ii. Metallic chlorates, when heated, decompose to form metal chloride and oxygen gas, e.g.,

$$2KClO_3(s) \rightarrow 2KCl(s) + 3O_2(g)$$

iii. Metallic hydroxides, when heated, decompose to form the metal oxide and water, e.g.,

$$Ca(OH)_2(s) \rightarrow CaO(s) + H_2O(l)$$

iv. Metallic carbonates, when heated, decompose to form the metal oxide and carbon dioxide, e.g.,

$$Li_2CO_3(s) \rightarrow Li_2O(s) + CO_2(g)$$

v. Metallic oxides are often stable, but a few decompose when heated to form the metal and oxygen gas, e.g.,

$$2HgO(s) \rightarrow 2Hg(l) + O_2(g)$$

4. Electrical decomposition (using electricity to cause decomposition—*electrolysis*).

i. Molten compounds will decompose into their elements via a REDOX reaction, e.g.,

$$2NaCl(l) \rightarrow 2Na(s) + Cl_2(g)$$

ii. Aqueous solutions will also undergo electrolysis, but water is sometimes oxidized or reduced rather than the elements that make up the solute; e.g., the electrolysis of *aqueous* potassium bromide yields bromine and hydrogen gas (from the water) (see Chapter 13).

$$2K^+(aq) + 2Br^-(aq) + 2H_2O(l) \rightarrow$$
$$Br_2(l) + 2K^+(aq) + H_2(g) + 2OH^-(aq)$$

C. Acid–base Reactions

1. Arrhenius acid is a substance that contains hydrogen and releases hydrogen ions (H^+); it is limited to aqueous solutions:

$$H_xB \rightarrow xH^+ + B^{x-}$$

2. Arrhenius base is a substance that releases hydroxide ions (OH^-) when placed in water; it is limited to aqueous solutions:

$$X(OH)_y \rightarrow X^{y+} + yOH^-$$

3. Neutralization reactions is when an Arrhenius acid and base react to produce a salt and water:

$$HCl(aq) + NaOH(aq) \rightarrow H_2O(l) + NaCl(aq)$$

$$H^+(aq) + OH^-(aq) \rightarrow H_2O(l) \text{ (the net ionic equation)}$$

4. Brønsted-Lowry acid is a species (neutral compound or cation or anion) that can donate a proton (H^+); it is the proton donor:

$$H_3PO_4(aq) + H_2O(l) \rightarrow H_2PO_4^-(aq) + H_3O^+(aq)$$

   In this reaction, the acid H_3PO_4 is donating a proton to the base, water.

5. Brønsted-Lowry base is a species (neutral compound or cation or anion) that can accept a proton; it is the proton acceptor:

$$NH_3(aq) + H_2O(l) \rightarrow NH_4^+(aq) + OH^-(aq)$$

   In this reaction, the base NH_3 is accepting a proton (H^+) from the acid, water.

6. Monoprotic acids
   i. Can only donate one proton per molecule or formula unit.
   ii. Examples include HCl, HF, and CH_3COOH.

7. Polyprotic acids
   i. Can donate more than one proton per molecule or formula unit.
   ii. Examples include H_2SO_4, H_3PO_4, and $H_2C_2O_4$ (the second or third H^+ may or may not be lost with similar ease to the first).
   iii. Anions of the examples above are SO_4^{2-}, PO_4^{3-}, and $C_2O_4^{2-}$.
8. Amphiprotic substances
   i. Can act as either an acid or a base.
   ii. Examples include H_2O, HCO_3^-, and $H_2PO_4^-$.

   Water is one of the most common examples.

$$H_2O(l) + H_2O(l) \rightleftharpoons H_3O^+(aq) + OH^-(aq)$$

   One water molecule donates H^+ and one accepts H^+.
9. Conjugate acid–base pairs
   i. A pair of substances that differ in formula only by a proton (H^+).
   ii. Conjugate acid is the partner of the Brønsted-Lowry base.
   iii. Conjugate base is the partner of the Brønsted-Lowry acid.
   iv. All acid–base reactions involving proton transfer have two conjugate acid–base pairs.
   v. Strong acids have correspondingly weak, conjugate bases and vice versa. In the example below, since HCl is a strong acid and Cl^- is a weak base:

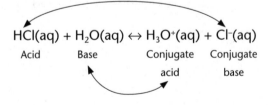

$$\underset{\text{Acid}}{HCl(aq)} + \underset{\text{Base}}{H_2O(aq)} \leftrightarrow \underset{\text{Conjugate acid}}{H_3O^+(aq)} + \underset{\text{Conjugate base}}{Cl^-(aq)}$$

10. Using the Brønsted-Lowry model

   i. The strength of an acid and a base can be determined by the percent of ionization.

   ii. HCl is a strong Brønsted-Lowry acid since it ionizes 100% to produce H_3O^+ and Cl^- in water. The molar concentrations of H_3O^+ and Cl^- are equal to the molar concentration of HCl. For example, if the concentration of HCl is 0.25 M, then the concentrations of H_3O^+ and Cl^- are 0.25 M since the reaction below goes 100% to the product side.

   $$HCl(aq) + H_2O(l) \rightarrow H_3O^+(aq) + Cl^-(aq)$$

0.25 M	0.25M	0.25M

   iii. NH_3 is a weak base and does not ionize completely in water. Hence, the concentration of OH^- is much less than the initial concentration of NH_3 since the reaction below only slightly goes to the product side.

   $$NH_3(g) + H_2O(l) \rightleftharpoons NH_4^+(aq) + OH^-(aq)$$

   iv. The Brønsted-Lowry model predicts that the stronger the acid, the weaker the conjugate base and vice versa.

   $$HCl(aq) + H_2O(l) \leftrightarrow H_3O^+(aq) + Cl^-(aq)$$

Acid	Base	Conjugate acid	Conjugate base

   HCl is a stronger acid than H_3O^+ because it is better in donating H^+, and the reaction effectively lies 100% to the product side.

11. Strong acids completely ionize in water, forming hydronium or the hydrogen ion, and are strong electrolytes since many ions are present in the solution.

12. Strong bases completely ionize in water, forming the hydroxide ion, and are strong electrolytes since many ions are present in the solution.

13. Weak acids and weak bases do not completely ionize in water and are considered weak electrolytes since very few ions are present in the solution.

Strong Acids	Strong Bases	Weak Acids	Weak Bases
HCl Hydrochloric acid	LiOH Lithium hydroxide	HCOOH Methanoic (formic) acid	NH_3 Ammonia
HBr Hydrobromic acid	NaOH Sodium hydroxide	CH_3COOH Ethanoic (acetic) acid	R-NH_2 Amines (R is a carbon chain)
HI Hydroiodic acid	KOH Potassium hydroxide	H_3PO_4 Phosphoric acid	Pyridines (organic compounds with N atoms)
HNO_3 Nitric acid	Group 2 hydroxides such as $Ba(OH)_2$ (barium hydroxide) and $Sr(OH)_2$ (strontium hydroxide)*	H_2CO_3 Carbonic acid	
H_2SO_4 Sulfuric acid			
$HClO_4$ Perchloric acid			

* Some group 2 hydroxides do not dissolve to any great extent in water, but 100% of any dissolved base will be ionized, and, as such, they can be considered strong bases.

Test Tip

The concepts of Lewis acids and Lewis bases are no longer tested on the AP Chemistry exam.

D. Oxidation-Reduction (REDOX) Reactions

1. Oxidation states (oxidation numbers) are the charges that an atom *actually* has when in an ionic compound, or when in a molecular compound, the charge that the atom *would* have if the compound were ionic.
2. Oxidation numbers are used to determine oxidized and reduced species in REDOX reactions.
3. Rules for assigning oxidation numbers:
    - i. Any element when uncombined (as the free element) is 0 (e.g., Cl_2, Fe, and S_8).
    - ii. The sum of the oxidation numbers in a neutral substance is 0.
    - iii. In an ion, the sum of the oxidation numbers equals the charge.
    - iv. Group 1 always +1, group 2 always +2, F always –1, O almost always –2, and H almost always +1.
    - v. In a binary compound with metals, the group 17 elements are –1, group 16 elements are –2, and group 15 elements are –3.
4. The role of the electron in oxidation-reduction (REDOX) reactions:
    - i. All REDOX reactions involve the transfer of electrons.
    - ii. If a substance accepts an electron, it is *reduced.*
    - iii. If a substance loses an electron, it is *oxidized.*
    - iv. During REDOX, electrons are transferred from the species that is oxidized to the species that is reduced.

Test Tip

Mnemonics to remember for REDOX reactions:

OILRIG—Oxidation Is Loss; Reduction Is Gain
"LEO the lion GERS"—Losing Electron Oxidation; Gaining Electron Reduction

    - v. A reducing agent is the species that reduces another species. In doing so, it becomes oxidized. The reducing agent loses electrons.

vi. An oxidizing agent is the species that oxidizes another species. In doing so, it becomes reduced. The oxidizing agent gains electrons.

vii. In order to determine the oxidizing and reducing agents, write the net ionic equation.

$$2Ag^{+}(aq) + Cu(s) \rightarrow 2Ag(s) + Cu^{2+}(aq)$$

Ag^{+} accepts electrons from Cu and is reduced to Ag.

Ag^{+} is the oxidizing agent.

Cu donates electrons to Ag^{+} and is oxidized to Cu^{2+}.

Cu is the reducing agent.

Test Tip

You will no longer see the terms *reducing agent* and *oxidizing agent* on the AP Chemistry exam. You will only need to understand REDOX in terms of electron transfer. However, it is likely that these terms will remain crucial to your understanding of REDOX chemistry, and you will almost certainly continue to encounter them.

5. Balancing REDOX reactions

   i. Write the half-reactions—one for the reduction and one for the oxidation.

   ii. Balance atoms other than oxygen and hydrogen for each half-reaction.

   iii. In acidic conditions, balance oxygen with H_2O and hydrogen with H^{+}. In basic conditions, balance oxygen with H_2O and hydrogen with H^{+}, *then* add OH^{-} to each side to neutralize H^{+} and cancel any water molecules.

   iv. Balance charge for each half-reaction by using electrons.

   v. If needed, multiply each half-reaction by a coefficient to balance electrons.

   vi. Add (merge) the half-reactions to cancel electrons and produce the full REDOX reaction.

Test Tip

Oxidation and reduction must take place together. They do not occur in isolation from one another. There has to be a "source" and a "sink" for electrons.

6. Examples of REDOX equations
   i. Simple REDOX (no acid or base considerations):

$Cr^{2+}(aq) + I_2(aq) \rightarrow Cr^{3+}(aq) + I^-(aq)$

Oxidation: $Cr^{2+}(aq) \rightarrow Cr^{3+}(aq)$

Balance atoms: $Cr^{2+}(aq) \rightarrow Cr^{3+}(aq)$

Balance charge: $Cr^{2+}(aq) \rightarrow Cr^{3+}(aq) + e^-$

Reduction: $I_2(aq) \rightarrow I^-(aq)$

Balance atoms: $I_2(aq) \rightarrow 2I^-(aq)$

Balance charge: $2e^- + I_2(aq) \rightarrow 2I^-(aq)$

Multiply the oxidation half-reaction by 2 in order to make the number of electrons the same in each half-reaction, and merge with the reduction half-reaction to cancel the electrons and give the full REDOX equation.

$2Cr^{2+}(aq) \rightarrow 2Cr^{3+}(aq) + 2e^-$

$2e^- + I_2(aq) \rightarrow 2I^-(aq)$

Full REDOX: $2Cr^{2+}(aq)^- + I_2(aq) \rightarrow 2Cr^{3+} + 2I^-(aq)$

   ii. Balancing in acidic solution:

$NO_3^-(aq) + Ag(s) \rightarrow NO_2(g) + Ag^+(aq)$

Oxidation: $Ag(s) \rightarrow Ag^+(aq)$

Balance atoms: $Ag(s) \rightarrow Ag^+(aq)$

Balance charge: $Ag(s) \rightarrow Ag^+(aq) + e^-$

Reduction: $NO_3^-(aq) \rightarrow NO_2(g)$

Balance atoms (using H_2O and H^+):
$2H^+_{(aq)} + NO_3^-(aq) \rightarrow NO_2(g) + H_2O(l)$

Balance charge: $e^- + 2H^+ + NO_3^-(aq) \rightarrow NO_2(g) + H_2O(l)$

No multiplication is required since the electrons are the same in each half-reaction, so simply merge the oxidation half-reaction with the reduction half-reaction to cancel the electrons and give the full REDOX equation.

$Ag(s) \rightarrow Ag^+(aq) + e^-$

$e^- + 2H^+ + NO_3^-(aq) \rightarrow NO_2(g) + H_2O(l)$

Full REDOX: $2H^+ + NO_3^-(aq) + Ag(s) \rightarrow NO_2(g) + H_2O(l) + Ag^+(aq)$

iii. Balancing in basic solution:

$MnO_4^-(aq) + I^-(aq) \rightarrow MnO_2(s) + IO_3^-(aq)$

Oxidation: $I^-(aq) \rightarrow IO_3^-(aq)$

Balance atoms (using H_2O and H^+): $3H_2O(l) + I^-(aq) \rightarrow IO_3^-(aq) + 6H^+(aq)$

Add OH^- (i.e., balance in basic solution by adding sufficient OH^- to each side in order to remove H^+ and make H_2O):

$6OH^-(aq) + 3H_2O(l) + I^-(aq) \rightarrow IO_3^-(aq) + 6H^+(aq) + 6OH^-(aq)$

Cancel the water: $6OH^-(aq) + I^-(aq) \rightarrow IO_3^-(aq) + 3H_2O(l)$

Balance charge: $6OH^-(aq) + I^-(aq) \rightarrow IO_3^-(aq) + 3H_2O(l) + 6e^-$

Reduction: $MnO_4^-(aq) \rightarrow MnO_2(s)$

Balance atoms: $4H^+(aq) + MnO_4^-(aq) \rightarrow MnO_2(s) + 2H_2O(l)$

Add OH^- (i.e., balance in basic solution by adding sufficient OH^- to each side in order to remove H^+ and make H_2O):

$4OH^-(aq) + 4H^+(aq) + MnO_4^-(aq) \rightarrow MnO_2(s) + 2H_2O(l) + 4OH^-(aq)$

Cancel the water: $2H_2O(l) + MnO_4^-(aq) \rightarrow MnO_2(s) + 4OH^-(aq)$

Balance charge: $3e^- + 2H_2O(l) + MnO_4^-(aq) \rightarrow MnO_2(s) + 4OH^-(aq)$

Multiply the reduction half-reaction by 2 in order to make the number of electrons the same in each half-reaction, and merge with the oxidation half-reaction to cancel the electrons and give the full REDOX equation.

$6e^- + 4H_2O(l) + 2MnO_4^-(aq) \rightarrow 2MnO_2(s) + 8OH^-(aq)$

$6OH^-(aq) + I^-(aq) \rightarrow IO_3^-(aq) + 3H_2O(l) + 6e^-$

Full REDOX: $H_2O(l) + 2MnO_4^-(aq) + I^-(aq) \rightarrow 2MnO_2(s) + IO_3^-(aq) + 2OH^-(aq)$

7. Important REDOX reactions

   i. Combustion of hydrocarbons to produce carbon dioxide and H_2O and liberate energy.

   $CH_4(g) + 2O_2(g) \rightarrow CO_2(g) + 2H_2O(g) + \text{energy}$

   ii. Metabolism of food (sugars, fats, and proteins), e.g., glucose during respiration.

   $C_6H_{12}O_6(g) + 6O_2(g) \rightarrow 6CO_2(g) + 6H_2O(g) + \text{energy}$

Practice Questions

1. A student heats a known mass of pure calcium carbonate in an open test tube, until no more carbon dioxide is produced. The sample in the tube had an original mass of 10.00 g and a final mass of 5.60 g.
   a. Write an equation to show the reaction that takes place.
   b. Classify the reaction.
   c. How would the student ensure that the reaction was complete?
   d. What mass of carbon dioxide has been produced in the reaction?
   e. What is the fundamental, chemical and scientific principle that allows the calculation in d. to be performed?
   f. Carbon dioxide *always* contains what ratio by mass of C and O?

2. In the following chemical reaction, determine the two conjugate acid–base pairs.

$$C_5H_5N(aq) + CH_3COOH(aq) \rightleftharpoons CH_3COO^-(aq) + C_5H_5NH^+(aq)$$

3. What is the conjugate base of H_2S?

4. What is the conjugate acid of CO_3^{2-}?

5. Determine the species that is oxidized and the species that is reduced in the following reaction:

$$CuSO_4(aq) + 2Na(s) \rightarrow Na_2SO_4(aq) + Cu(s)$$

6. In acid solution, manganese (VII) ions are reduced to Mn^{2+}. In a REDOX titration, 21.00 mL of 0.1500 M $KMnO_4$ is required to oxidize a 25.00 mL sample of Fe^{2+} ions.
   a. Calculate the moles of MnO_4^- used in the titration.

b. Construct the two half-reactions in the titration and, hence, the full REDOX reaction.

c. Calculate the concentration of Fe^{2+} in the solution.

Answers

1. a. $CaCO_3(s) \rightarrow CaO(s) + CO_2(g)$

   b. Decomposition

   c. By repeatedly heating, cooling, and weighing until a constant mass is achieved and thereby knowing that all of the carbon dioxide has been liberated

   d. 10.00 g – 5.60 g = 4.40 g

   e. The law of conservation of mass

   f. 12.01 g of C : 32.00 g of O

2. C_5H_5N (base) and $C_5H_5NH^+$ (conjugate acid)

   CH_3COOH (acid) and CH_3COO^-(conjugate base)

3. HS^- (H_2S must be the acid and lose a proton.)

4. HCO_3^- (CO_3^{2-} must be the base and gain a proton.)

5. The net ionic equation is:

   $$Cu^{2+}(aq) + 2Na(s) \rightarrow 2Na^+(aq) + Cu(s)$$

   Cu^{2+} is being reduced since it gains electrons to form Cu; Na is being oxidized since it loses electrons to form Na^+.

6. a. (0.02100 L)(0.1500 mol/L) = 3.150×10^{-3}mols

   b. $MnO_4^- + 8H^+ + 5e^- \rightarrow Mn^{2+} + 4H_2O$

   $$Fe^{2+} \rightarrow Fe^{3+} + e^-$$

   $$MnO_4^- + 8H^+ + 5Fe^{2+} \rightarrow Mn^{2+} + 4H_2O + 5Fe^{3+}$$

c. Moles of Fe^{2+} = (5)(mols of MnO_4^-) = (5)(3.150×10^{-3} mols) = 1.575×10^{-2} mols

$$\text{Conc. of } Fe^{2+} = \frac{(\text{mols of } Fe^{2+})}{(\text{vol of } Fe^{2+})}$$

$$= \frac{(1.575 \times 10^{-2} \text{ mols})}{0.02500 \text{ L}}$$

$$= 0.6300 \text{ M}$$

Chapter 13

Energy Changes in Chemical Reactions

I. Driving Forces and Observations

A. Driving Forces—Why Do Reactions Occur?

1. Formation of a precipitate
2. Formation of a gas
3. Transfer of electrons (REDOX reaction)
4. Formation of water from H^+ and OH^- (acid–base reaction)

B. Observations—Help to Identify Physical and Chemical Change

1. Seeing a color change
2. Seeing a solid (precipitate) being formed
3. Observing an energy change in the form of light or heat

C. Physical Versus Chemical Change

1. Physical changes involve only the making and breaking of intermolecular forces *between* substances, and *no* new substances are formed.
2. Chemical changes involve the making and breaking of intra bonds (ionic or covalent) *within* substances, and new substances *are* formed.
3. Using only macroscopic observations, it can sometimes be difficult to distinguish between physical and chemical changes. By considering the forces or bonds that are being broken, the type of change can be identified.

II. Energy Changes

A. Endothermic and Exothermic Reactions

1. Exothermic reactions release energy to the surroundings, and the surroundings increase in temperature, e.g., the combustion of fuels.
2. Endothermic reactions absorb energy from the surroundings, and the surroundings decrease in temperature.
3. Diagramming endothermic and exothermic change.
   i. As energy transfers to and from surroundings

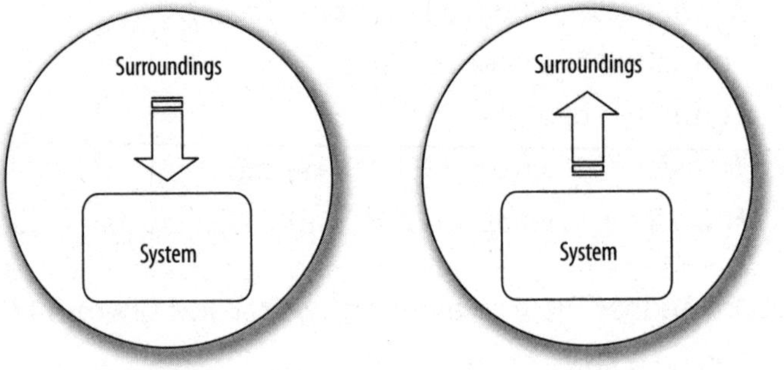

Endothermic, $\Delta H = +$ Exothermic, $\Delta H = -$

   ii. In energy diagrams (profiles)

Endothermic Reaction Exothermic Reaction

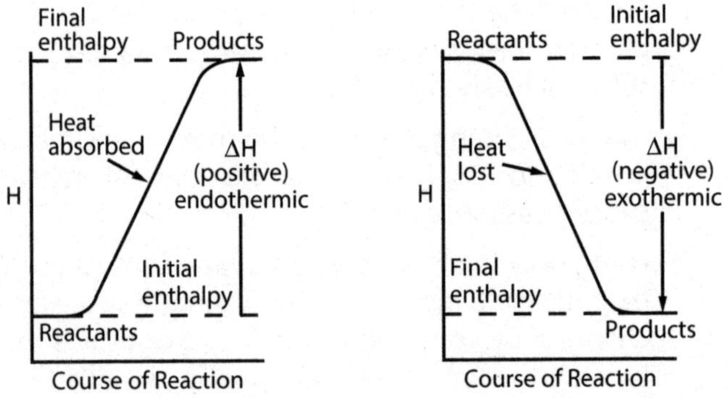

III. Electrochemistry

A. Galvanic Cells

1. Galvanic cells use thermodynamically-favored REDOX reactions to produce electrical energy via a flow of electrons (also known as Voltaic cells or batteries).
2. Salt bridge allows ions to flow, completing the circuit in the galvanic cell.
3. Anode—oxidation will occur, thus producing electrons (it will be the half-cell with the more negative standard electrode reduction potential).
4. Cathode—reduction will occur, thus gaining electrons (it will be the half-cell with the more positive standard electrode reduction potential).
5. Standard electrode reduction potential/E° is the measure of the reduction potential of an individual half-cell when compared to the standard hydrogen electrode. The standard hydrogen electrode is given a value of 0.00 V. All values are measured under standard conditions of 1.0 M solutions, 298K, and gases at 1 atm.

   Positive standard electrode reduction potential values mean that the species in question will be reduced more readily than H^+ ions and vice versa.

   The voltage of a Galvanic cell is calculated by applying:

   $$E^\circ_{cell} = E^\circ_{Reduced} - E^\circ_{Oxidized}$$

   Do NOT change the signs of the standard electrode reduction potentials if applying this method. However, an alternative method (that yields the same answer) *does* involve the reversal of the sign of the standard electrode reduction potential of the oxidized species, followed by the *addition* of the standard electrode reduction potential of the reduced species. Do NOT mix the two methods. Either keep the standard electrode reduction potentials as they are given (both as reductions) and subtract the oxidized form from the reduced form, *or* reverse the sign of the standard electrode reduction potential of the oxidized species and add.

Example: Consider the REDOX reaction shown below and the standard electrode reduction potentials that follow:

$$Cu^{2+}_{(aq)} + Zn_{(s)} \rightarrow Zn^{2+}_{(aq)} + Cu_{(s)}$$

$$Cu^{2+}_{(aq)} + 2e^{-} \rightleftharpoons Cu_{(s)} \quad +0.34 \text{ V}$$

$$Zn^{2+}_{(aq)} + 2e^{-} \rightleftharpoons Zn_{(s)} \quad -0.76 \text{ V}$$

Either realize that copper is reduced and zinc is oxidized and apply:

$$E^{\circ}_{cell} = E^{\circ}_{Reduced} - E^{\circ}_{Oxidized}$$

$$E^{\circ}_{cell} = +0.34 - -0.76 = 1.10 \text{ V}$$

or

Realize that the zinc is oxidized, so reverse *its* standard electrode reduction potential and then add.

$$E^{\circ}_{cell} = +0.34 + (+0.76) = 1.10 \text{ V}$$

6. Gibbs Free Energy relates to E° via the following equation:

$$\Delta G^{\circ}_{rxn} = -nF E^{\circ}$$

n = number of moles of electrons transferred

F = Faraday constant = 96485 Coulombs per mole of electrons

E° = standard cell voltage

If E° equals a negative number, then ΔG°_{rxn} equals a positive number and the reaction is not favored

If E° equals a positive number, then ΔG°_{rxn} equals a negative number and the reaction is favored

Test Tip

If you calculate a galvanic cell voltage as a negative value, then it is the reverse reaction (the positive voltage) that is the favored REDOX reaction. Essentially, you've written the spontaneous chemical reaction backwards!

7. The representation of galvanic cells
   i. A "cell diagram" is not a "picture" of the apparatus but a nomenclature convention that always shows oxidized species on the left and reduced species on the right separated by a double line to represent the salt bridge. For example, $Zn|Zn^{2+}||Cu^{2+}|Cu$
   ii. A "picture" of the cell shows the oxidized and reduced species, as well as the salt bridge and flow of electrons. The "picture" may or may not have the oxidized half-cell on the left and the reduced half-cell on the right—take care to check and do not assume.

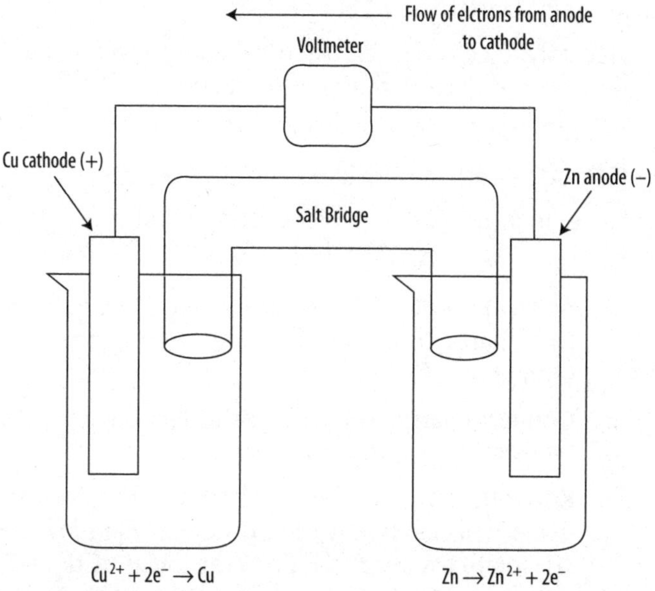

8. Nonstandard conditions
   i. All voltages that are calculated using the standard electrode reduction potentials assume standard conditions. Sometimes, these conditions are not in effect.
   ii. When nonstandard conditions are in effect, apply Le Chatelier's principle (see Chapter 24) to see if the reaction will shift forward or backward to achieve standard conditions.

iii. Reactions that shift forward have higher voltages than standard; reactions that shift backwards have lower voltages than standard.

9. Concentration cells

   i. A concentration cell is made from two half-cells with the same metal/ion combinations, but with differing ion concentrations.

   ii. Electrons flow from the lower concentration half-cell to the higher concentration half-cell in order to achieve equilibrium.

B. Electrolytic (Electrolysis) Cells

1. Electrolytic cells require electrical energy to force a nonfavored REDOX reaction to occur.
2. Anode—oxidation will occur.
3. Cathode—reduction will occur.
4. Have negative $E°$ and positive ΔG°_{rxn} values.
5. Quantitative aspects of electrolysis

   i. Apply $q = I\,t$ to calculate the charge in Coulombs (q = charge in Coulombs; I = current in amps, t = time in seconds)

   ii. Convert charge in Coulombs to Faradays, by dividing q by 96485 coulombs per Faraday.

   iii. Know that the number of electrons that appear in a given half-reaction is the number of Faradays required to complete the reaction. For example, in $Cu^{2+} + 2e^{-} \rightarrow Cu$, one mole (63.5 g) of copper is formed by the passage of 2 Faradays of electrical charge and masses produced are proportional. For example, passing 1 Faraday in the copper half-reaction results in the formation of 0.5 moles of copper metal.

Test Tip

A table of standard reduction potentials is no longer given on the AP Chemistry exam. Any values that you need will be included in the questions.

Practice Questions

1. Assume that you are constructing a Galvanic cell based on the half-reactions between $Zn^{2+}(aq)/Zn(s)$ and $Ag^{+}(aq)/Ag(s)$. The standard reduction potentials are given below.

   $Ag^{+}(aq) + e^{-} \rightarrow Ag(s)$ $\qquad E° = 0.80$ V

   $Zn^{2+}(aq) + 2e^{-} \rightarrow Zn(s)$ $\qquad E° = -0.76$ V

   a. Write a balanced chemical REDOX equation and determine $E°$ for the cell.

   b. What will be observed at the cathode and at the anode?

   c. Is the reaction thermodynamically favored? Explain your answer.

2. The Downs cell is an industrial cell that is used to produce sodium metal and chlorine gas by the electrolysis of molten sodium chloride.

   a. Write the half-reaction that you would expect to occur at the anode.

   b. Write the half-reaction that you would expect to occur at the cathode.

   c. Write the full REDOX reaction that takes place in the cell.

   d. A Downs cell is run for 1.00 hour with a current of 20.0 amps. How many moles of chlorine gas would be produced in the cell?

Answers

1. a. Based on the standard electrode potentials, the silver half-cell will undergo reduction, while the zinc half-cell will undergo oxidation.

   Multiply the silver equation by 2 and merge it with the zinc equation in order to cancel electrons.

   Reduction: $2Ag^{+}(aq) + 2e^{-} \rightarrow 2Ag(s)$

   Oxidation: $Zn(s) \rightarrow Zn^{2+}(aq) + 2e^{-}$

   ==============================

   $2Ag^{+}(aq) + Zn(s) \rightarrow Zn^{2+}(aq) + 2Ag(s)$

   $E^{\circ}_{cell} = 0.80\text{ V} - (-\,0.76\text{ V}) = 1.56\text{ V}$

   b. At the cathode, silver ions are reduced to form silver metal: $2Ag^{+}(aq) + 2e^{-} \rightarrow 2Ag(s)$

   At the anode, the zinc electrode dissolves to form aqueous zinc ions: $Zn(s) \rightarrow Zn^{2+}(aq) + 2e^{-}$

   c. Since the sign for E° is positive, the forward reaction has a negative ΔG° value, and it is thermodynamically favored.

2. a. $2Cl^{-} \rightarrow Cl_2 + 2e^{-}$

   b. $Na^{+} + e^{-} \rightarrow Na$

   c. $2NaCl \rightarrow 2Na + Cl_2$

   d. $q = (20)(60 \times 60 \times 1) = 7.20 \times 10^{4}\text{ c}$

   $$\frac{7.20 \times 10^{4}\text{ c}}{96485\text{ c per Faraday}} = 0.746\text{ F}$$

   $$\frac{0.746\text{ F}}{2\text{ F}} = \frac{x\text{ mol } Cl_2}{1\text{ mol } Cl_2}$$

   $x = 0.373$ moles of Cl_2.

PART V

RATES OF REACTION

Chapter 14

Factors Affecting Rates of Reaction

> ***Note:* It is likely that you will have been taught the contents of Chapter 15 *before* the contents of Chapter 14, and that makes sense. This book is compiled in the order of the Big Ideas and Enduring Understandings outlined in the College Board course and exam description.**

I. Kinetics

A. Concept of Rate of Reaction

1. Rate of a chemical reaction is a measure of the change in concentration of reactants or products over time.
2. Can be measured as the *decrease* of the reactant concentration per unit time.
3. Can be measured as the *increase* of the product concentration per unit time.
4. One important method employed to measure rate is using Beer's law to determine the concentration of a colored solution as a reaction proceeds.

B. Conditions That Can Affect Rate

1. Increasing the surface area of a solid reactant can increase the rate by increasing the number of collisions between the reactant particles (see Chapter 15).
2. Catalyst increases the rate by lowering the activation energy of a reaction (see Chapter 17).

3. Increasing the temperature results in a faster reaction. The rate constant is temperature dependent and a rise in temperature will increase the rate constant (see below).
4. Concentration of reactants increases the amount of reactants colliding with each other, thus yielding product (see Chapter 15).

C. Use of Experimental (Concentration) Data and Graphical Analysis to Determine Reactant Order, Rate Laws, and Rate Constants

1. General formula for rate equation. For the generic reaction

$$aA + bB + cC \rightarrow dD$$

Rate of reaction = $k[A]^x [B]^y [C]^z$

where k is the rate constant and x, y, z are the orders with respect to A, B, and C, respectively, but are *not* necessarily the stoichiometric coefficients of A, B, and C.

2. Orders

i. The order with respect to a reactant is the exponent of the concentration term in the rate equation (a.k.a. the *rate law*).

$$2H_2O_2(aq) \rightarrow 2H_2O(l) + O_2(g)$$

Rate of reaction = $k[H_2O_2]$

In this case, the reaction is *first order* with respect to the reactant H_2O_2 since the concentration of H_2O_2 is raised to the power of 1 in the rate equation.

ii. Zero order rate reaction

➤ The rate of the reaction is independent of the concentration of the reactant(s).

$$aA \rightarrow bB$$

$$\text{Rate} = k[A]^0 = k$$

iii. First order rate reaction

- The rate of the reaction is directly proportional to the concentration of one of the reactants.

$$aA \rightarrow bB$$

$$\text{Rate} = k[A]^1$$

(sometimes the "1" is omitted; i.e., Rate = $k[A]$)

- Radioactive decay reactions are a very common example of a first-order process. They have a constant half-life, meaning the time taken for half of the atoms to decay is constant and independent of the initial concentration.

For these reactions, $k = \dfrac{0.693}{\text{half-life}}$.

iv. Second order rate reaction

- The rate of the reaction is directly proportional to the square of the concentration of one of the reactants.

$$aA \rightarrow bB$$

$$\text{Rate} = k[A]^2$$

v. Determining orders

Conduct multiple experiments with changing concentrations of each reactant and measure the rate in some way.

$$2NO(g) + Cl_2(g) \rightarrow 2NOCl(g)$$

Trial	[NO] mol/L	[Cl_2] mol/L	Rate
1	0.200	0.200	1.20×10^{-6}
2	0.400	0.200	4.80×10^{-6}
3	0.200	0.400	2.40×10^{-6}
4	0.400	0.400	9.6×10^{-6}

- Comparing Trial 1 and Trial 2—the concentration of NO is doubled, and the concentration of Cl_2 remains the same. The reaction rate is increased 4 times, meaning the rate of reaction with respect to NO is second order ($2^2 = 4$).
- Comparing Trial 1 and Trial 3—the concentration of Cl_2 is doubled, and the concentration of NO stays the same. The reaction rate is doubled, meaning the rate of reaction with respect to Cl_2 is first order ($2^1 = 2$).
- Comparing Trial 1 and Trial 4—both reactants are doubled; therefore, the reaction rate is increased 8 times ($2^3 = 8$).

Rate = $k[NO]^2[Cl_2]$

The reaction is second order with respect to NO, first order with respect to Cl_2 and third order overall. *The overall order is the sum of the individual orders.*

The magnitudes of rate constants vary a lot, since reactions can vary from being very slow to very fast, and they have varying units. The units of rate constants are commonly required on AP questions.

vi. Using graphs to find orders—Different orders require different plots to generate straight lines.

Order	Rate Law	Straight Line Plot	Derived From
0	Rate = k	[*A*] versus T	–
1	Rate = $k[A]$	*In*[*A*] versus T	$\ln[A]_t - \ln[A]_0 = -kt$
2	Rate = $k[A]^2$	$\frac{1}{[A]}$ versus T	$\frac{1}{[A]_t} - \frac{1}{[A]_o} = kt$

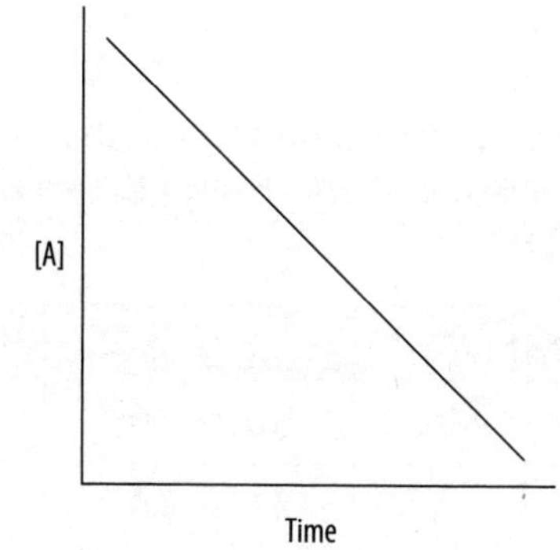

If the best fit line follows the data to the left, then the reaction is zero order.

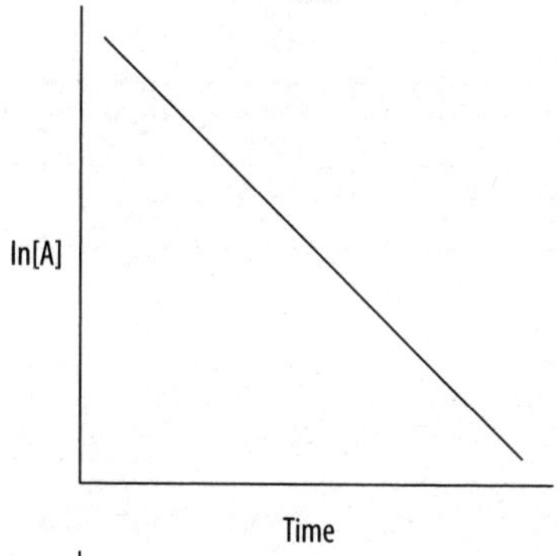

If the best fit line follows the data to the left, then the reaction is first order.

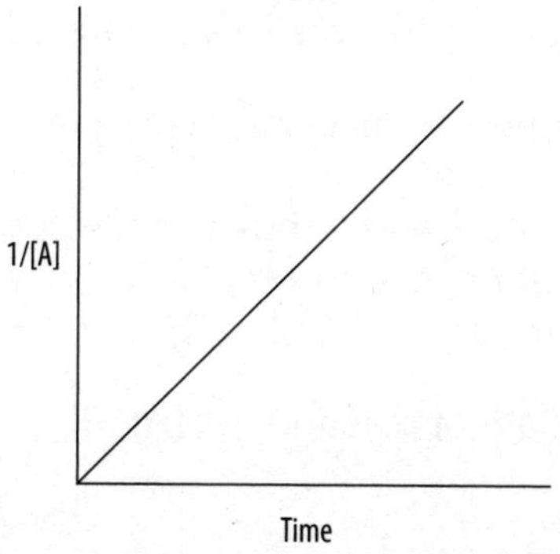

If the best fit line follows the data to the left, then the reaction is second order.

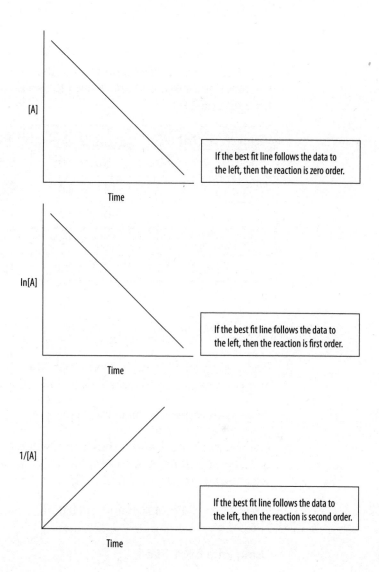

Practice Question

1. Based on the data below, determine the rate equation for the reaction between *A* and *B*, and the value and units of the rate constant.

Trial	[*A*] M	[*B*] M	Rate Ms^{-1}
1	2.1×10^{-3}	1.2×10^{-3}	5.60×10^{-6}
2	2.1×10^{-3}	2.4×10^{-3}	11.2×10^{-6}
3	2.1×10^{-3}	0.6×10^{-3}	2.80×10^{-6}
4	4.2×10^{-3}	1.2×10^{-3}	11.2×10^{-6}
5	6.3×10^{-3}	1.2×10^{-3}	16.8×10^{-6}

Answer

1. Comparing Trials 1 and 3 shows that when [*B*] is halved and [*A*] is constant, the rate halves. This indicates first order with respect to reactant *B* $\left(\left(\frac{1}{2}\right)^1 = \frac{1}{2}\right)$.

   Comparing Trials 1 and 5 shows that when [*A*] is tripled and [*B*] is constant, the rate triples. This indicates first order with respect to reactant A ($3^1 = 3$).

   $$\text{Rate of reaction} = k\,[A]\,[B]$$

   Using data from Trial 1:

   $5.60 \times 10^{-6}\ Ms^{-1} = k\,(2.1 \times 10^{-3}\,M)\,(1.2 \times 10^{-3}\,M)$

   $$k = 2.2\ M^{-1}\ s^{-1}$$

Collision Theory

I. Collision Theory

A. Elementary Reactions

1. Those reactions that proceed from reactants to products in a single step with a single transition state are called elementary reactions.

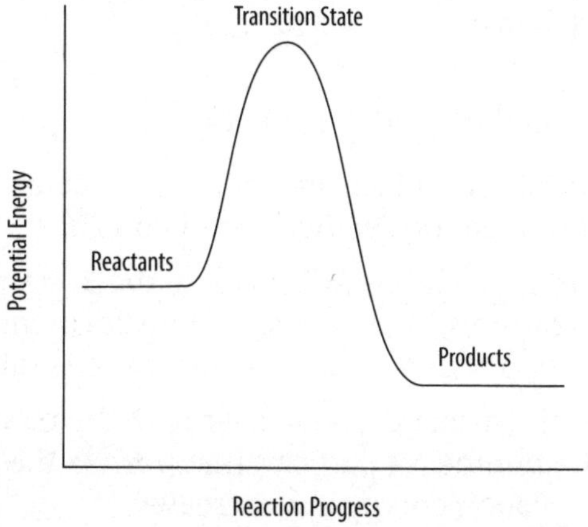

2. Elementary reactions involving a single reactant are called *unimolecular.*
3. Elementary reactions involving two reactants are called *bimolecular.*
4. Elementary reactions involving three or more reactant particles colliding are extremely rare.

B. Collision Theory—Fundamental Conditions for Reactions to Occur

1. Reacting molecules must collide with each other.
2. Reacting molecules must collide with sufficient energy (*activation energy*). Activation energy is an energy barrier that must be overcome before a reaction can take place.
3. Reacting molecules must collide with a specific geometry or orientation to allow the rearrangement of reactant bonds into product bonds.
4. "Successful" or "effective" collisions are those that have *both* the sufficient energy *and* the correct orientation to allow them to lead to a reaction.
5. "Unsuccessful" collisions are those that lack sufficient energy and/or correct orientation, and, therefore, they do *not* lead to a reaction.
6. In most reactions, only a small percentage of the collisions are "effective."

C. Factors Related to Collision Theory

1. Increasing the surface area of a solid reactant can increase the rate by increasing the number of collisions.
2. A rise in temperature will increase the average kinetic energy of the particles, thus allowing more collisions between reactants. The net result is to increase the rate of the reaction.
   i. As the average kinetic energy of the particles increases, the number of particles that possess the required activation energy also increases.

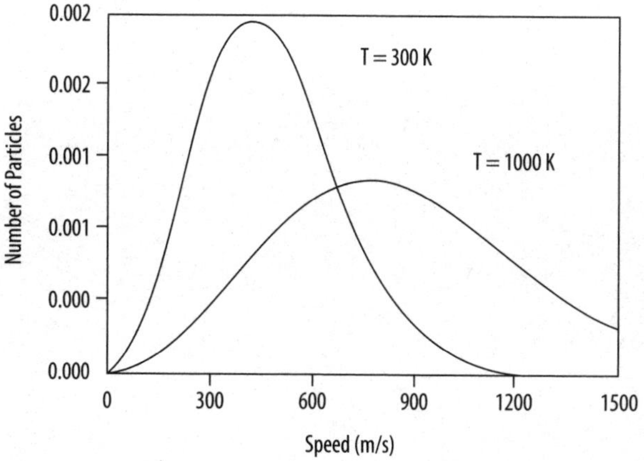

D. Arrhenius Equation

1. The Arrhenius Equation relates the *rate constant* for a chemical reaction to the temperature and the activation energy of the reaction.
2. The rate constant, k, is related to temperature (T) and activation energy (Ea), thus,

$$k = Ae^{(-Ea/RT)}$$

k *increases* with increasing temperature.

k *decreases* with increasing activation energy.

Chapter 16

Mechanisms

I. Reaction Mechanisms

A. Mechanisms and Elementary Steps

1. Reaction mechanism is the sequence of bond-breaking and bond-making steps that lead to an overall chemical reaction. Each individual step is called an elementary step.
2. If one reactant is involved in an elementary step, it is called unimolecular.
3. If two reactants are involved in an elementary step, it is called bimolecular.
4. Elementary steps involving two or more reactants are extremely rare, since that would require three particles to simultaneously collide.
5. The rate of reaction for an elementary step is based on the stoichiometric coefficients of the reactants in the elementary step; however, this should not be confused with using the stoichiometric coefficients of the *overall* reaction made up from a series of elementary steps to determine the order. That method is *incorrect* and must not be used.*

B. The Relationship Between the Rate-Determining Step in a Mechanism and the Rate Law

1. The slowest elementary step is called the rate-determining step (RDS).

* It would be *incorrect* to use the stoichiometry of the *overall* reaction to infer that the reaction were second order with respect to [NO], first order with respect to [F_2], and third order overall. To determine the orders with respect to individual reactants and hence the overall order, use ONLY the RDS.

2. The rate of the reaction is dependent upon the rate-determining step. For example, the overall reaction, $2NO_2 + F_2 \rightleftharpoons 2NO_2F$, has a two-step mechanism.

   Step 1: $NO_2 + F_2 \rightleftharpoons NO_2F + F$ (slow, RDS)

   Step 2: $NO_2 + F \rightleftharpoons NO_2F$ (fast)

   In the slow, rate-determining step 1, the rate equation is Rate = $k\,[NO_2][F_2]$, i.e., it is a second order reaction (1 + 1 = 2).

   Experimentally, it is found that the reaction is second order overall, which is consistent with the slow rate-determining step also being second order.

   In short, the overall rate equation and the RDS must be in agreement in terms of both the species present in each and in terms of matching the stoichiometry with the orders.

3. The individual steps of the mechanism must sum to give the reaction stoichiometry of the overall reaction.

4. Intermediates that appear in rate laws are generally replaced by the reactants they depend upon. For example, if, in the reaction, $A + 2B \rightarrow C + D$ the following mechanism is known:

Step 1:	$A \rightleftharpoons Q$	(fast)
Step 2:	$Q + B \rightarrow D$	(slow, RDS)
Step 3:	$B \rightarrow C$	(fast)

   The rate equation for the slow step, i.e., the rate equation for the reaction overall is:

   $$\text{Rate} = k\,[Q][B]$$

   However, since Q is an intermediate, it can be replaced by the reactant that it depends upon in the earlier, fast equilibrium step, i.e., A.

   With A replacing Q, the rate equation becomes

   $$\text{Rate} = k\,[A][B]$$

Practice Question

1. Consider the proposed mechanism for a reaction between a number of gases, shown below.

Step 1: $NO_2(g) + CO(g) \rightarrow NO(g) + CO_2(g)$

Step 2: $NO(g) + O_3(g) \rightarrow NO_2(g) + O_2(g)$

   a. Identify the catalyst in the reaction. Explain your answer.

   b. Identify the intermediates in the reaction. Explain your answer.

   c. Write the overall reaction.

Answer

1. a. $NO_{2(g)}$. It is present at the beginning of the reaction, then changes, but is ultimately regenerated and, therefore, present at the end of the reaction.

   b. $NO_{(g)}$. It is generated during step 1 and then used up in step 2. It is not present at the beginning or the end of the reaction.

   c. $CO_{(g)} + O_{3(g)} \rightarrow CO_{2(g)} + O_{2(g)}$

Catalysts

I. Catalysts

A. Concept of a Catalyst

1. Increases the rate of a reaction by lowering the activation energy of an elementary step in a reaction, but leaves the mechanism unchanged, or
2. Increases the rate of a reaction by forming a new reaction intermediate and, therefore, a new reaction mechanism that happens to have a lower activation energy.

B. Diagrammatic Representations of Catalysts

1. In an energy profile,

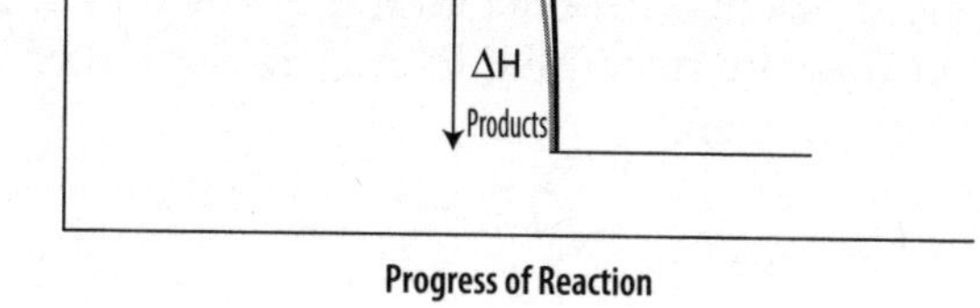

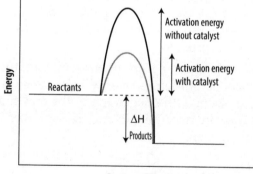

Lower activation energy means that a greater number of particles will possess the required minimum energy (E_{act}) and a greater number of collisions will be successful, i.e., they will lead to a reaction.

2. In a Maxwell-Boltzman distribution,

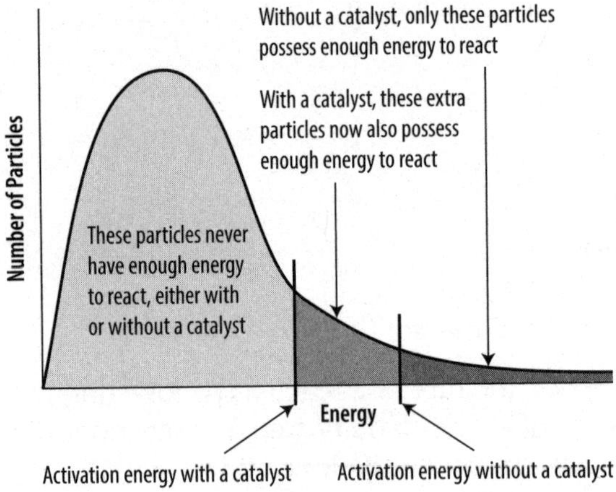
Without a catalyst, only these particles possess enough energy to react
With a catalyst, these extra particles now also possess enough energy to react
Number of Particles
These particles never have enough energy to react, either with or without a catalyst
Energy
Activation energy with a catalyst
Activation energy without a catalyst

C. Recognizing a Catalyst in a Reaction Mechanism

1. Catalysts are present at the start of a reaction *and* at the end of a reaction, i.e., they are not consumed during the reaction.
2. Sometimes catalysts are described as "not taking part in the reaction"; however, that can be a little misleading. They *do* often react and change during a reaction, but they ultimately change back into their original format, are regenerated, and so are not consumed.
3. When combining the steps of a mechanism, the catalyst will cancel out. (Intermediates will also cancel out, but they are not present at the beginning or the end of a reaction; rather they are generated in one step and then used up in a subsequent step.)

D. Types of Catalysts

1. Acid–base—A proton (H^+) is lost or gained by a reactant, which results in a change of the rate of reaction.
2. Surface—For example, gases may be adsorbed onto the surface of a metal where reactant bonds are weakened

and, as a result, the reaction occurs faster. Also, collecting reactants onto a common surface in a smaller area effectively increases reactant concentration, which in turn means an increase in rate (see Chapters 14 and 15).

3. Enzyme—Biological catalysts (often proteins) with active sites that interact with substrates (reactants) to increase the rate of reaction.

Practice Question

1. Identify the catalyst in this reaction mechanism.

$$O_3 + Cl \rightarrow O_2 + ClO$$

$$ClO + O \rightarrow Cl + O_2$$

Answer

1. Cl is the catalyst since it is present at the beginning of the reaction and at the end of the reaction, but when the steps of the mechanism are merged to produce the full, chemical reaction, it cancels out.

Test Tip

Don't confuse catalysts with intermediates. Intermediates are generated during a reaction and are used up in a subsequent step, so they also cancel out. They are not present at the beginning of a reaction or at the end of a reaction. In the practice question above, ClO is an intermediate.

PART VI

CHEMICAL THERMODYNAMICS

Chapter 18

Temperature, Heat, and Specific Heat Capacity

I. Temperature

A. Temperature and Kinetic Energy

1. Temperature—the measure of the average kinetic energy of the particles of a substance.
   i. The Kelvin temperature is proportional to the average kinetic energy. For example, doubling the Kelvin temperature doubles the average kinetic energy.
   ii. As absolute zero is approached (0 K), the particles approach zero kinetic energy.
   iii. A Maxwell-Boltzmann distribution shows how the particles at a high temperature have greater kinetic energies than those at a low temperature.

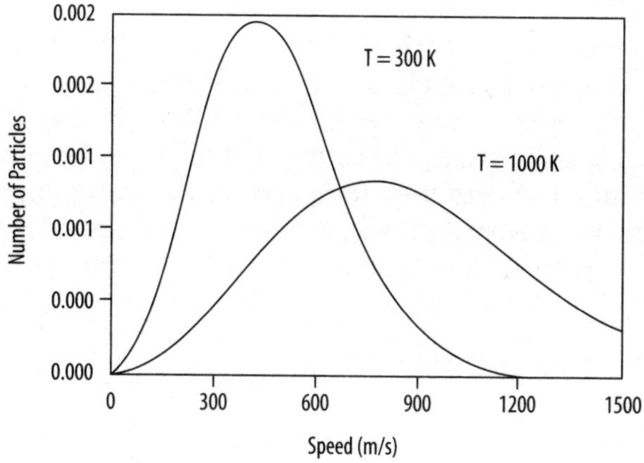

iv. Particle diagrams with vectors can be used to illustrate differing kinetic energies at differing temperatures.

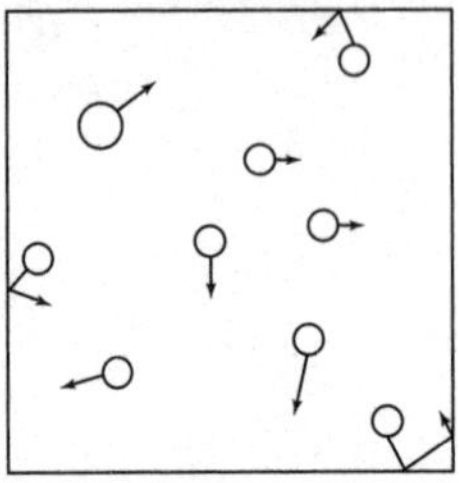

Low Temperature

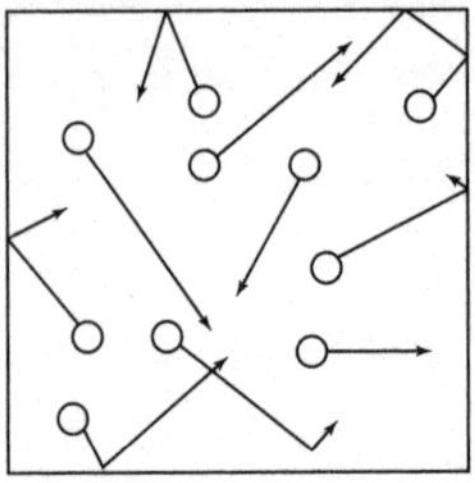

High Temperature

II. Heat

A. When kinetic energy is transferred from one object to another, that energy transferred is called heat.

1. When molecules that have different temperatures collide, kinetic energy is transferred from the warmer one to the cooler one until the temperatures (average kinetic energies) are the same.
2. When the average kinetic energies are the same, *thermal equilibrium* is reached.
3. Specific heat capacity is most easily thought of as the heat (energy) required to raise the temperature of 1 g of a substance by one degree Celsius. Metals have relatively low specific heats, indicating that relatively less energy is required to raise their temperatures. By comparison, water has a relatively high specific heat, so will require much more energy to achieve a similar temperature change.

$$\text{Specific Heat Capacity } (C_p) = \frac{\text{quantity of heat supplied}}{\text{(mass of object)(temperature change)}}$$

Substance	Specific Heat (J/g · K)
Al	0.902
H_2O (l)	4.184
Glass	0.84

4. Measuring the amount of heat transferred can be calculated using this equation:

$$q = (m)(c)(\Delta T)$$

where q = heat transferred
m = mass of substance
c = specific heat capacity
$\Delta T = T_{final} - T_{initial}$ = change in temperature

ΔT Object	Sign of ΔT	Sign of *q*	Direction of Heat Transfer
Increase	+	+	Heat transferred into object
Decrease	–	–	Heat transferred out of object

Practice Question

1. A 21.3 g sample of a metal is heated to 70.0°C and dropped into 62.4 g of the water at 21.0°C. The final temperature of the water is 24.0°C. Given that the specific heat capacity of water is 4.184 $Jg^{-1}{}^{\circ}C^{-1}$, calculate the specific heat of the metal.

Answer

1. $q_{water} = q = (m)(c)(\Delta T)$

   $q_{water} = (62.4\text{ g})(4.184\ \frac{J}{g}\cdot{}^{\circ}C)(24.0^{\circ}C - 21.0^{\circ}C)$

   $q_{water} = 783\text{ J}$

   Assuming that no energy is lost, and since heat was transferred from the metal to the water:

   $q_{water} = -q_{metal}$

   $-q_{metal} = q = (m)(c)(\Delta T)$

   $--783\text{ J} = (21.3\text{ g})(c)(24.0^{\circ}C - 70.0^{\circ}C)$

   $c = 0.799\frac{J}{g}\cdot{}^{\circ}C$

Chapter 19

Conservation of Energy, Work, and Calorimetry

I. Work

A. Gas Expansion in a Cylinder and Piston

1. As a gas expands inside a cylinder, its particles will collide with a piston and energy will be transferred from the gas to the piston. As a result, the piston moves.
2. When energy is transferred in this way, the gas is said to be "doing work" on the piston.

B. Conservation of Energy

1. The work done in the cylinder and piston system can be calculated using $P\Delta V$, where P is the external pressure that the gas is working against, and ΔV is the change in volume of the gas.
2. As work is done, the gas loses energy and the piston gains energy. Energy must be conserved, so the energy lost by the gas is equal in magnitude to the energy gained by the piston.

II. Heating and Cooling Curves

A. Heating Curves—What Are They?

1. A plot of temperature against time (at constant pressure) that results when heating a solid substance, through its melting and boiling points, until it becomes a hot gas.

B. Interpreting the Heating Curve

1. Starting with a solid below its melting point, the following is observed:

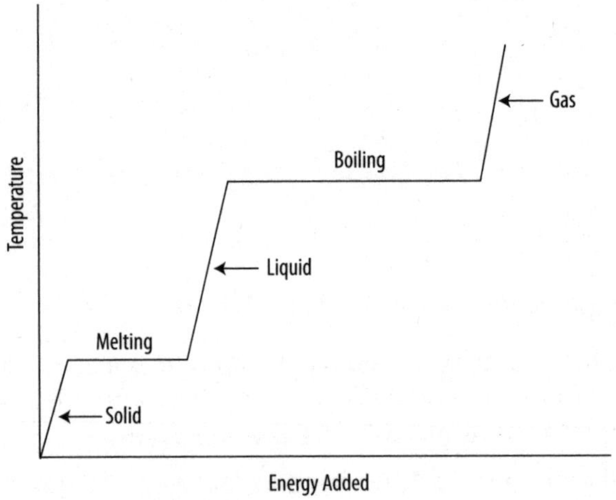

i. The temperature of the solid increases at a constant rate until it begins to melt.

ii. When melting (fusion) begins, the temperature remains constant until the entire solid has been converted to a liquid.

iii. The temperature of the liquid increases at a constant rate until it begins to boil.

iv. When boiling (vaporization) begins, the temperature remains constant until the entire liquid has been converted to a gas.

v. The temperature of the gas increases at a constant rate.

vi. In the regions where the temperature of the solid, liquid, or gas is being increased, the amount of energy that is added can be calculated using $q = m\,c\,\Delta T$, where m is mass, c is the specific heat capacity of the substance being heated, and ΔT is the temperature change.

Note: Specific heat capacities for different phases of the same substance are different; for example, ice, water, and steam all have different specific heat capacities, despite all being H_2O.

vii. A plateau region represents a stage where two phases are in equilibrium with one another.

- *Standard enthalpy of fusion* is the energy change when 1 mole of a substance is converted from a solid to a liquid. Where the solid is melting, the amount of energy that is added can be calculated using $q = (\Delta H_{fusion})$(moles).
- *Standard enthalpy of vaporization* is the energy change when 1 mole of a substance is converted from a liquid to a gas. Where the liquid is boiling, the amount of energy that is added can be calculated using $q = \Delta H_{vaporization})$ (moles)
- These are endothermic processes because energy must be added.

Test Tip

Here is a helpful way to understand the process: the energy being added is either being used to increase temperature OR is being used to affect a phase change—not both.

C. Cooling Curves—What Are They?

1. Cooling curves are plots of temperature against time (at constant pressure) that result when cooling a gaseous substance, through its condensation and freezing points, until it becomes a cold solid.

D. Interpreting the Cooling Curve

1. Starting with a gas above its boiling point, the following is observed:

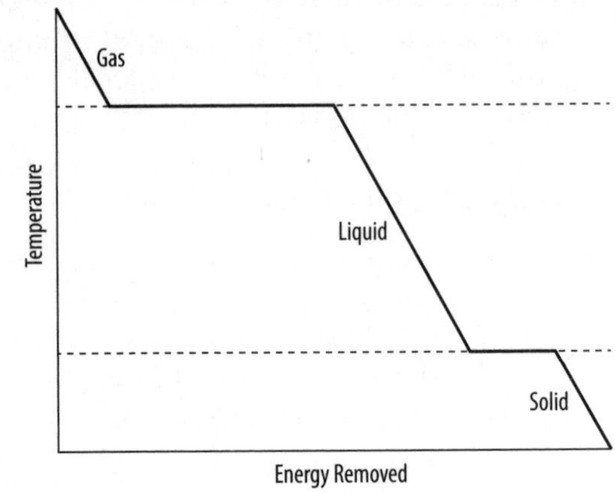

i. The temperature of the gas decreases at a constant rate until it begins to condense.

ii. When condensing begins, the temperature remains constant until all the gas has turned to liquid.

iii. The temperature of the liquid decreases at a constant rate until it begins to freeze.

iv. When freezing begins, the temperature remains constant until all the liquid has turned to a solid.

v. The temperature of the solid decreases at a constant rate.

vi. In the regions where the temperature of the solid, liquid, or gas is being decreased, the amount of energy that is removed can be calculated using $q = m\, c\, \Delta T$, where m is mass, c is the specific heat capacity of the substance being heated, and ΔT is the temperature change.

vii. A plateau region represents a stage where two phases are in equilibrium with one another.

- Where the gas is condensing, the amount of energy that is removed can be calculated using $q = (\Delta H_{vaporization})(\text{moles})$.
- Where the liquid is freezing, the amount of energy that is removed can be calculated using $q = (\Delta H_{fusion})(\text{moles})$.
- These are exothermic processes because energy must be removed.

III. Calorimetry

A. Calorimetry is an experimental technique used to measure the change in energy of a chemical reaction or phase change.

1. General principles: Put a chemical reaction or phase change in contact with a heat bath (usually water). We can measure the change in temperature of the heat bath. Knowing the heat capacity of the heat bath, we can calculate the energy change in the heat bath by applying:

$$q = (m)(c)(\Delta T)$$

The energy change in the heat bath will be the same magnitude in energy as the chemical reaction or phase change, just with the opposite sign. If the heat bath gains energy, its temperature goes up, meaning the energy of the chemical reaction or phase change went down (it lost energy) and vice versa.

Test Tip

Enthalpy changes where energy is released are exothermic and have negative values.

Enthalpy changes where energy is absorbed are endothermic and have positive values.

2. Coffee cup calorimeter—Styrofoam cups are commonly used as insulators in the high school chemistry lab, to measure temperature changes without a loss of energy to the surroundings.

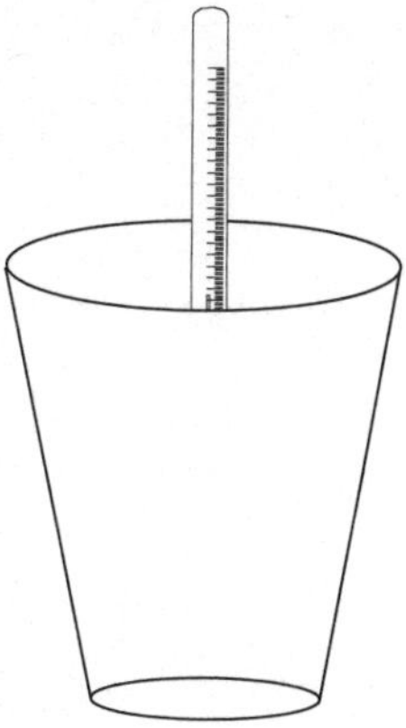

Practice Question

1. If 5.00 g of urea are added to 90.00 g of water in a coffee cup calorimeter, the temperature of the contents of the cup falls by 3.100 °C. If the specific heat capacity of the solution is 4.184 J/g°C, calculate the energy change in the coffee cup.

Answer

1. $q = m\,c\,\Delta T$

   $q = (95.00\text{ g})\ (4.184\text{ J/g°C})\ (3.100\text{°C})$

   $q = +1232\text{ J}$ (endothermic)

Chapter 20

Bond Making and Bond Breaking

I. Chemical Bonds

A. Nature of a Covalent Bond

1. The negative electrons of one atom and the positive nucleus of another attract one another.
2. If the nuclei of two atoms get too close to one another, their like charges repel.
3. The bond length between atoms in a covalent bond is the distance between the centers of the atoms, when the potential energy of the electrostatic (Coulombic) interaction is at a minimum.

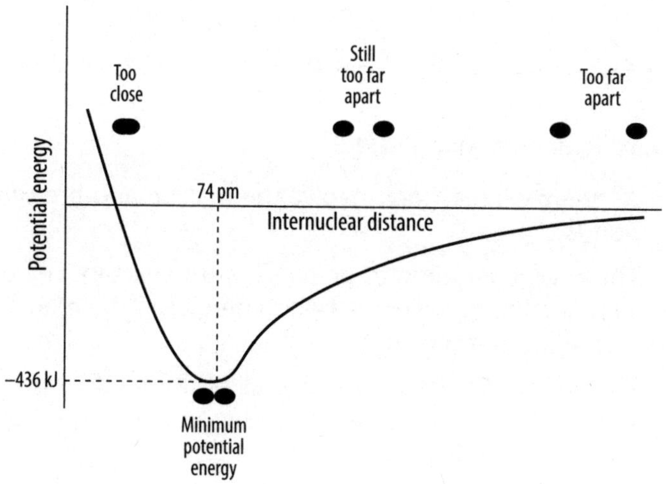

4. Since the bond-making process is an electrostatic one, it is governed by Coulomb's law.
   i. Double and triple bonds are stronger than single bonds, since the double and triple bonds contain a greater number of electrons than single bonds and the Coulombic attractions are stronger.
   ii. Shorter bonds, with atoms closer to one another, tend to be stronger than longer bonds with atoms further apart, since the distance between the charges is smaller.
5. Breaking bonds is an endothermic process that requires energy (endothermic changes are given positive values).
6. Making bonds is an exothermic process that releases energy (exothermic changes are given negative values).
7. Breaking or making a particular bond involves the same magnitude of energy change, i.e., the energy change is the same for breaking and making a particular bond, all that differs is the sign. For example, an average bond energy of 347 kJ for the C–C bond, means that 347 kJ of energy will be released (–347) when the bond is broken in a reaction, and that 347 kJ of energy is required to break the bond (+347) in a reaction.

II. Bond Energies in Chemical Reactions

A. Breaking and Making Bonds

1. All chemical reactions involve the making and breaking of bonds.
2. The energy required to break all of the reactant bonds is a sum of all of the average bond energies in the reactants. This is an endothermic term.
3. The energy released in making all of the product bonds is a sum of all of the average bond energies in the products. This is an exothermic term.

4. The energy change in the reaction as a whole is the sum of the breaking (endothermic) process and the making (exothermic) process, and can be negative or positive depending on the relative magnitude of the two (breaking and making) processes.
5. In an exothermic reaction, thermal energy is transferred *to* the surroundings. In an endothermic reaction, thermal energy is transferred *from* the surroundings. In each case the energy transfer is called the enthalpy change or ΔH, and it is usually measured in kJ/mol.

- *System* refers to a particular part of the universe that is being studied, and, in chemistry, this usually simply means a certain, chemical reaction.
- *Surroundings* is everything that is outside of the system.
- *Universe* is the system AND the surroundings.
- *Endothermic* is energy transferred from the surroundings to the system.
- *Exothermic* is energy transferred from the system to the surroundings.

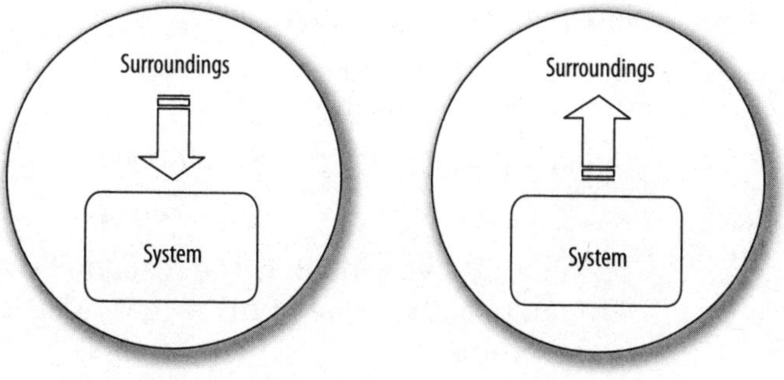

Endothermic, $\Delta H = +$

Exothermic, $\Delta H = -$

- *Standard Enthalpies of Formation* is the energy change when 1 mole of a substance is formed from its elements in their standard states. Also called $\Delta H°_f$.

➤ *Standard Enthalpy of a Reaction* is the energy change when a reaction occurs, with all products and reactants in a standard state. Sometimes called ΔH°_{rxn}. Superscript $^\circ$ indicates standard conditions of 1 bar (1 atm = 1.01325 bar) and although not necessarily true, often indicates 25°C as well.

Enthalpy Change for a Reaction = $\Delta H^\circ_{rxn} = \sum[\Delta H^\circ_f\,(\text{products})] - \sum[\Delta H^\circ_f\,(\text{reactants})]$

Test Tip

Remember, when an element is formed from itself, there is no change. The standard enthalpy of formation of elements in their standard states is zero.

III. Hess's Law

A. Manipulating Chemical Equations

1. *Hess's Law* states that the overall enthalpy change in a reaction is the sum of all the reactions for the process and is independent of the route taken.

   i. Rule 1: If you reverse the reactions, then change the sign of ΔH. For example,

   $A + B \rightarrow C$ $\quad\quad \Delta H = 60$ kJ

   $C \rightarrow A + B$ $\quad\quad \Delta H = -60$ kJ

   ii. Rule 2: If you multiply the reaction by a coefficient, multiply the value of ΔH by the same coefficient. For example,

   $A + B \rightarrow C$ $\quad\quad \Delta H = 60$ kJ

   $2A + 2B \rightarrow 2C$ $\quad\quad \Delta H = 120$ kJ

   iii. Rule 1 and 2 can be combined. For example, if the first reaction is tripled *and* reversed,

   $A + B \rightarrow C$ $\quad\quad \Delta H = 60$ kJ

   $3C \rightarrow 3A + 3B$ $\quad\quad \Delta H = -180$ kJ

Practice Questions

1. Based on the equation below, what is the enthalpy change when 56.4 g of $C_2H_5OH(l)$ decomposes?

$$C_2H_5OH(l) \rightarrow 2C(graphite) + 3H_2(g) + \tfrac{1}{2}O_2(g)$$
$$\Delta H = +\ 228\ kJ/mol$$

2. What is the standard enthalpy change, ΔH°_{rxn}, for the reaction below? The enthalpies of formation:

CH_4 – 75.0 kJ/mol, CO_2 – 394 kJ/mol, and H_2O – 286 kJ/mol

$$CH_4 + 2O_2 \rightarrow CO_2 + 2H_2O$$

3. Given the data below, calculate the enthalpy change for the decomposition of phosphorous trichloride.

Equation 1: $4PCl_3(g) \rightarrow P_4(s) + 6Cl_2(g)$ $\quad \Delta H = ?$

Equation 2: $P_4(s) + 10Cl_2(g) \rightarrow 4PCl_5(g)$ $\quad \Delta H = +\ 813.0\ kJ$

Equation 3: $PCl_3(g) + Cl_2(g) \rightarrow PCl_5(g)$ $\quad \Delta H = +\ 1146\ kJ$

Answers

1. $$\frac{56.4\ g\ C_2H_5OH}{1} \times \frac{1\ mol\ C_2H_5OH}{46.068\ g\ C_2H_5OH} \times \frac{228\ kJ}{1\ mol\ C_2H_5OH} = 280\ kJ$$

2. *Enthalpy Change for a Reaction* =

$$\Delta H^\circ_{rxn} = \sum\left[\Delta H^\circ_f\,(products)\right] - \sum\left[\Delta H^\circ_f\,(reactants)\right]$$

Enthalpy Change for a Reaction = ΔH°_{rxn}

$$\sum\left[2\ mole\ H_2O\left(-286\frac{kJ}{mol}\right) + 1\ mole\ CO_2\left(-394\frac{kJ}{mol}\right)\right] - \sum\left[1 mole\ CH_4\left(-75.0\frac{kJ}{mol}\right)\right]$$

$$= -891\ kJ$$

3. Reverse equation #2:

$$4PCl_5(g) \rightarrow P_4(s) + 10Cl_2(g) \qquad \Delta H = -813.0 \text{ kJ}$$

Multiply equation #3 by 4:

$$4PCl_3(g) + 4Cl_2(g) \rightarrow 4PCl_5(g) \qquad \Delta H = +\ 4584 \text{ kJ}$$

Sum the new equations and cancel out $4PCl_5(g)$ and $4Cl_2(g)$ from both sides:

$$4PCl_3(g) \rightarrow P_4(s) + 6Cl_2(g) \qquad \Delta H = +\ 3771 \text{ kJ}$$

Chapter 21

Inter and Intra Forces and Physical and Chemical Change

I. Chemical and Physical Change

A. Differences at the Particulate Level

1. When weak intermolecular forces (between molecules) are broken or formed, *physical* changes take place.
2. When strong intra bonds (chemical bonds *within* compounds) are broken or formed, *chemical* changes take place.
3. Some changes (e.g., dissolving an ionic salt in water) involve both intermolecular and chemical bonds changing, and, as such, can be classified as chemical and/or physical changes.
4. In large molecules (particularly those encountered in biochemistry), intermolecular forces can occur between different parts of the *same* molecule. When this occurs, molecules can change shape and, as a result, their properties can be affected. For example, enzymes and synthetic polymers have properties that are affected by their shape.

II. Intermolecular Forces

A. Intermolecular Forces of Attraction and Coulomb's Law

1. There is a Coulombic force of attraction between the opposite charges on different molecules that is governed by Coulomb's law.
2. These electrostatic attractions can take various forms: dipole–dipole, dipole–induced dipole, and induced dipole–induced dipole

B. Dipole–Dipole and Dipole–Induced Dipole

1. Dipole–dipole are Coulombic (electrostatic) attractions between permanent dipoles in adjacent molecules.
   i. Molecules with polar bonds (caused by differences in electronegativity between the atoms), and dipoles that do not cancel via symmetry (see Chapter 9), will have permanent dipoles.
   ii. When molecules that have permanent dipoles approach one another, they will arrange themselves so that the negative and the positive ends of the molecules attract one another.
   iii. Dipole–dipole interactions are intermediate in strength in terms of the (weak) intermolecular forces.
2. Dipole–induced dipole have the same Coulombic (electrostatic) interactions as dipole–dipole, except only one pair of molecules has a permanent dipole. The other is nonpolar, but has a dipole induced by the approach of the other polar molecule.
3. Hydrogen bonding occurs when hydrogen atoms, bonded to certain elements, create unusually large dipole–dipole, intermolecular attractions.
   i. Hydrogen is an exceptional element in that when it forms a covalent bond its electron is held predominately to one side of the nucleus leaving the other side almost completely exposed.
   ii. Any approaching negatively charged group can get very close to the hydrogen nucleus and this produces an unusually large Coulombic attraction.
   iii. These electrostatic attractions are exaggerated when H is bonded to a more electronegative element that is small enough to allow a significant intermolecular interaction, i.e., fluorine, oxygen, or nitrogen. Such exaggerated intermolecular electrostatic attractions are called *hydrogen bonds*.
   iv. The occurrence of hydrogen bonds has two important consequences.

➤ It gives substances containing them anomalously high boiling points.

Hydrogen Halide	Normal Boiling Point, °C
HF*	19
HCl	–85
HBr	–67
HI	–35

*High BP attributed to hydrogen bonding

➤ Substances containing them tend to have increased viscosity.

Both consequences are explained by the increased attraction between molecules caused by hydrogen bonding, making it more difficult to separate them.

v. Hydrogen bonds can occur between different parts of a single molecule. For example, in biology, the secondary and (in part) the tertiary structures formed by proteins are produced by internal hydrogen-bonding interactions between different parts of a single protein molecule.

vi. Hydrogen bonds are the strongest of the (weak) intermolecular forces.

C. Induced Dipole–Induced Dipole (London Dispersion Forces)

1. London dispersion forces are small Coulombic (electrostatic) forces that are caused by the movement of electrons within the covalent bonds of molecules that would otherwise have no permanent dipole.
2. As one molecule approaches another, the electrons of one or both are temporarily displaced owing to their mutual repulsion. This movement causes small, temporary induced dipoles to be set up which attract one another. These attractions are called *London dispersion forces (LDFs).*

3. These dispersion forces increase with an increasing number of electrons in the molecule and with an increasing surface area. This leads to more dispersion forces, greater attraction, and, therefore, higher melting and boiling points in molecules with more electrons and greater surface areas.

Test Tip

In the past, AP examiners have been keen for you to use the word "polarizable" to describe the increased ability of large molecules to form extensive LDFs; at the same time, they have been equally keen for you to not *reference increased mass as a reason for increasing LDFs.*

4. London dispersion forces exist between *all* molecules.
5. London dispersion forces are the weakest of the (weak) intermolecular forces.

Test Tip

It should be noted that although LDFs are the weakest of the intermolecular forces that you will study in AP Chemistry, with larger molecules the cumulative attraction of many *LDFs can sometimes outweigh a stronger intermolecular force such as a dipole–dipole interaction. For example, Cl_2 (only LDFs) has a higher boiling point than HCl (dipole–dipole and LDFs), since the much larger Cl_2 molecules have sufficient, cumulative LDFs to outweigh the combination of LDFs and dipole-dipole interactions between molecules of HCl.*

Entropy and Free Energy

Chapter 22

I. Entropy

A. Second Law of Thermodynamics

1. *Second Law of Thermodynamics* states that entropy or disorder of the universe will increase over time.
2. *Entropy* is the measurement of disorder given the symbol S and usually has the units $JK^{-1}mol^{-1}$
   i. Entropy increases with dispersal of particles.
   ii. Entropies of gases are larger than liquids, and liquid entropies are larger than solids, as the volume and dispersal of particles increases from solid to liquid to gas.
   iii. Entropies are greater for more complex molecules, are greater when the number of individual particles of the same phase increases, are greater with increased temperature (increased kinetic energy), and are greater when a single gas occupies a larger volume.
   iv. Since the entropy of a perfect, pure crystal at 0 K is given a value of zero, all absolute entropies for individual substances above 0 K, i.e., in the real world, are positive.
   v. The entropy change for a system (reaction) is calculated from the absolute entropies of the products and reactants.

   $$\Delta S^o_{system} = \sum[S°(\text{products})] - \sum[S°(\text{reactants})]$$

   If $\Delta S°_{system}$ is negative, then decrease in entropy (disorder).

   If $\Delta S°_{system}$ is positive, then increase in entropy (disorder).

II. Free Energy

A. Gibbs Free Energy

1. Standard Free Energies of Formation is the free energy change when 1 mole of a substance forms from its elements in their standard states, under standard conditions, and has the symbol $\Delta G°_f$. Elements have a value of zero, since forming an element from itself results in no change.
2. Standard Gibbs Free Energy Change for a reaction =

$$\Delta G^0_{reaction} = \sum[G^0_f(\text{products})] - \sum[G^0_f(\text{reactants})]$$

$$\Delta G°_{reaction} = \Delta H° - T\Delta S°$$

   i. If $\Delta G° > 0$, then the reaction is *not thermodynamically favored* and energy must be added to the reaction for it to occur. Reactant formation is favored.

   ii. If $\Delta G° < 0$, then the reaction *is thermodynamically favored* and no outside source of energy needs to be added. Product formation is favored.

   iii. If $\Delta G° = 0$, the reaction is at equilibrium. Reactant and product formation are equally favored.

The terms **spontaneous** *and* **non-spontaneous** *are commonly used in chemistry to describe thermodynamically favored and thermodynamically unfavored reactions, respectively. Look out for them in places* **other** *than on the AP exam, which is now using the terms* **favorable** *and* **unfavorable.**

3. *Gibbs Free Energy* is the amount of energy in a reaction that can be used for work.

4. The relationship between ΔH, ΔS, and ΔG

Sign of ΔH	Sign of ΔS	Sign of $-T\Delta S$	Sign of ΔG
(–)	(+)	(–)	– (always thermodynamically favored)
(–)	(–)	(+)	– at low temp (thermodynamically favored) + at high temp (not thermodynamically favored)
(+)	(+)	(–)	+ at low temp (not thermodynamically favored) – at high temp (thermodynamically favored)
(+)	(–)	(+)	+ (never thermodynamically favored)

i. Thermodynamically favored reactions may still not occur in any measurable rate if they are slow, i.e., if they have a high *activation energy*. When this happens the reaction is said to be under *kinetic control.*

ii. Reactions that are not thermodynamically favored can be forced to proceed by applying an external source of energy, such as electricity in electrolysis and battery charging, or the use of light and other electromagnetic radiation in photosynthesis and ionization processes.

iii. Combining thermodynamically unfavored reactions with thermodynamically favored reactions, via their common intermediates, can lead to a favorable thermodynamic process overall. This is called *coupling*.

5. Relationship of change in free energy to equilibrium constants—See Chapter 26.

Practice Question

1. Based on the table of values below for the following reaction at 298K, determine ΔH, ΔS, ΔG, and whether the reaction is thermodynamically favored or not.

$$H_2S(g) + 2O_2(g) \rightleftharpoons H_2SO_4(l)$$

	H_2S	O_2	H_2SO_4
ΔH_f° (kJ/mol)	–20	0	–814
ΔS° (J/K · mol)	206	205	157
ΔG_f° (kJ/mol)	–33	0	–690

Test Tip

Note the values of zero for the ΔH°_f and ΔG°_f for $O_{2(g)}$. It is an element in its standard state, so when it is formed from itself, there is no change.

Test Tip

Usually, units of enthalpy (H) are in terms of kJ and units of entropy (S) are in terms of J. You must convert them to the same units (kJ or J) before using the $\Delta G^\circ = \Delta H^\circ - T\Delta S^\circ$ expression.

Answer

1. $\Delta H^\circ = [1(-814)] - [1(-20)] = -794$ kJ mol^{-1}

   $\Delta S^\circ = [1(157)] - [1(206) + 2(205)] = -459$ J K^{-1} mol^{-1}

   $\Delta G^\circ = [1(-690)] - [1(-33)] = -657$ kJ mol^{-1}

Check using

$$\Delta G^\circ = \Delta H^\circ - T\Delta S^\circ = -794 - [298(-0.459)] = -657 \text{ kJmol}^{-1}$$

If ΔG is negative, then the product formation is favored.

PART VII

EQUILIBRIUM

Chapter 23

Dynamic Equilibrium

I. Equilibrium

A. Dynamic Equilibrium: Physical, Biological, Environmental, and Chemical

1. Many chemical reactions are reversible.
2. Many biological and environmental examples of equilibrium exist, such as oxygen binding to, and being released from, hemoglobin, and the carbon cycle.
3. Physical examples include the evaporation and condensation of H_2O.
4. Chemical examples include the exchange of H^+ ions in acid base reactions, and the exchange of electrons in REDOX reactions.
5. Double arrows indicate reversibility of the reaction ($\rightleftharpoons$).

B. Equilibrium Conditions

1. At equilibrium, both reactants and products are present.
2. The forward reaction is favored when the concentration of reactants is high, and, hence, the rate of conversion of reactants to products is high.
3. The backward reaction is favored when the concentration of products is high, and, hence, the rate of conversion of products to reactants is high.
4. Dynamic equilibrium is reached when the forward and backward reactions continue at the same rate, and the concentrations of the reactants and products are constant.

Test Tip

> **Constant** *concentration of reactants and products does not necessarily mean* **identical** *concentrations of reactants and products. In fact, they are very rarely the same numerical value.*

Test Tip

> *All concentrations being* **constant** *(not changing) is not the same thing as concentrations being* **equal** *(all having the same value). This is a common misconception and one that you must not have.*

5. Graphically, the concentrations of reactants and products can be represented as:

    i. For a product-favored reaction:

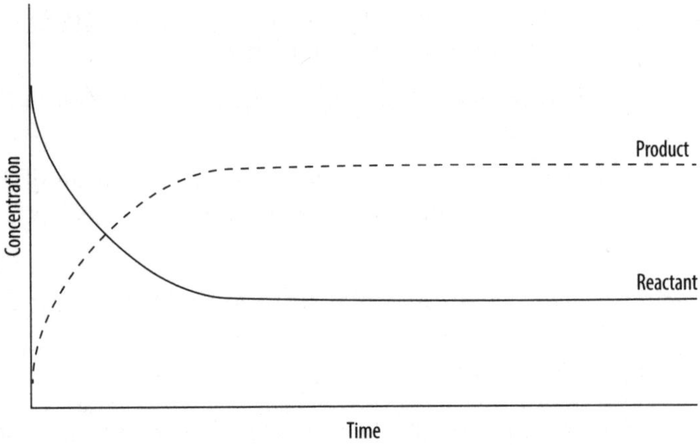

ii. For a reactant-favored reaction:

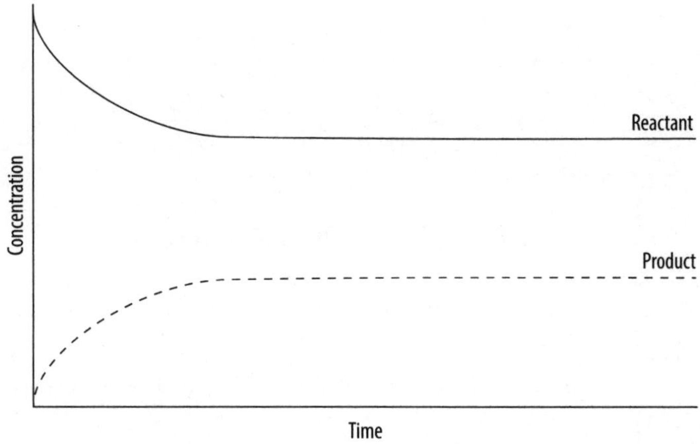

C. Quantitative Treatment

1. Equilibrium constants for gaseous reactions: *Kp, Kc*

i. Equilibrium constant relates the concentrations of reactants and products at equilibrium at a given temperature.

For the general reaction: $aA + bB \rightleftharpoons cC + dD$

The equilibrium constant = $K = \dfrac{[C]^c[D]^d}{[A]^a[B]^b}$

- ➤ Product molar concentrations are in the numerator.
- ➤ Reactant molar concentrations are in the denominator.
- ➤ Each concentration is raised to the power of its stoichiometric coefficient in the balanced chemical equation.
- ➤ Pure solids and pure liquids (e.g., water) are not placed into the expression.
- ➤ The expression *Kc* indicates that concentrations are used (moles per liter).
- ➤ The expression K_p indicates that partial pressures are used (pressure units, often atm) and *only* gases are included.

Examples

1. $N_2(g) + 3H_2(g) \rightleftharpoons 2NH_3(g)$

$$K_c = \frac{[NH_3]^2}{[N_2][H_2]^3} \qquad K_p = \frac{P_{NH_3}^2}{P_{N_2}P_{H_2}^3}$$

2. $2KClO_3\,(s) \rightleftharpoons 2KCl(s) + 3O_2(g)$

$$K_c = [O_2]^3 \qquad K_p = P_{O_2}^3$$

3. $NH_3(aq) + H_2O(l) \rightleftharpoons NH_4^+(aq) + OH^-(aq)$

$$K_c = \frac{[NH_4^+][OH^-]}{[NH_3]}$$

D. Manipulation of the Equilibrium Constant

1. Multiplying a chemical equation by a coefficient

   i. Example:

$$A(s) + \frac{1}{2}B_2(g) \rightleftharpoons AB(g) \qquad K = 2.3 \times 10^3$$

If the equation above is multiplied by 2, then the value of the new K is the original K squared:

$$2A(s) + B_2(g) \rightleftharpoons 2AB(g) \qquad K = 5.3 \times 10^6$$

Rule: If the balanced chemical equation is multiplied by a factor, then the new value of *K* is the original *K* raised to that power.

2. Reversing a chemical reaction

   i. Example:

      $A(s) + B(g) \rightleftharpoons AB(g) \qquad K = 4.0 \times 10^3$

      If the equation above is reversed, then the value of the new K is the reciprocal of the original K:

      $AB(g) \rightleftharpoons A(s) + B(g) \qquad K = 2.5 \times 10^{-4}$

      Rule: If the balanced chemical equation is reversed, then the new value of K can be expressed as follows:

      $K_{new} = 1/K_{original}$

3. When several equations are used to obtain a net balanced chemical equation.

   i. Example:

      Equation 1: $C\text{ (graphite)} + \tfrac{1}{2}\, O_2(g) \rightleftharpoons CO(g)$
      $K = 2.28 \times 10^{23}$

      Equation 2: $C\text{ (graphite)} + O_2(g) \rightleftharpoons CO_2(g)$
      $K = 1.25 \times 10^{69}$

      What is the equilibrium constant for the following reaction?

      Equation 3: $2CO(g) + O_2(g) \rightleftharpoons 2CO_2(g)$

      Manipulate equations 1 and 2;

      Equation 1: Reverse and multiply by 2, to give;
      Equation 4: $2CO(g) \rightleftharpoons 2C\text{ (graphite)} + O_2(g)$

      New K for reaction 4: must be reciprocal *and* square of original; $K = 1.92 \times 10^{-47}$

      Equation 2: Multiply by 2, to give;
      Equation 5: $2C\text{ (graphite)} + 2O_2(g) \rightleftharpoons 2CO_2(g)$

      K must be original K squared; $K = 1.56 \times 10^{138}$

Since equation 3 is a combination of equations 4 and 5, multiply K for equations 4 and 5 to get $K = 3.00 \times 10^{91}$.

Rule: If a balanced chemical equation is produced by the combination of other equations, then the new value of K can be expressed as follows:

$$K_{new} = K_1 \cdot K_2 \text{ etc.}$$

E. What Does the Equilibrium Constant Mean?

1. If $K > 1$, then the product formation is favored—equilibrium concentrations of products are large. In a particulate diagram, product particles will dominate reactant particles.
2. If $K < 1$, then the reactant formation is favored—equilibrium concentrations of reactants are large. In a particulate diagram, reactant particles will dominate product particles.
3. If $K = 1$, the formation of reactants and products are equally favored.

F. Reaction Quotient, Q

1. A measure of the proportions of the products to reactants at any point (*not necessarily equilibrium*) in a chemical reaction.
2. Not always measured at equilibrium, but can be. If it is, then $Q = K$.

$$aA + bB \rightarrow cC + dD$$

$$Q = \text{reaction quotient} = \frac{[C]^c[D]^d}{[A]^a[B]^b}$$

i. If $Q < K$, the system is not at equilibrium and product formation is favored.

ii. If $Q > K$, the system is not at equilibrium and the reactant formation is favored.

iii. If $Q = K$, the reaction is at equilibrium.

Practice Questions

1. Equilibrium concentrations are given (plug directly into *K* expression).

   A mixture of HI, H_2, and I_2 gases are in a reaction vessel and allowed to reach equilibrium at a temperature of 723K.

   $$2HI(g) \rightleftharpoons H_2(g) + I_2(g)$$

   The equilibrium concentrations for each are listed below. Calculate the K_c for the reaction at this temperature.

   $$[HI] = 2.50 \times 10^{-2}M$$
   $$[H_2] = 6.33 \times 10^{-4}M$$
   $$[I_2] = 1.23 \times 10^{-3}M$$

2. Equilibrium concentrations are not given (use the **I**nitial, **C**oncentration, **E**quilibrium table—**ICE** table).

   1.00 mol of H_2 and 1.00 mol of S_2 are placed in a 1.00L flask at 900K.

   $$2H_2(g) + S_2(g) \rightleftharpoons 2\ H_2S(g)$$

   When equilibrium is achieved, 0.750 mol of H_2S has formed. Calculate K_c at 900K for the reaction.

3. Working with partial pressures (K_p)

   Dinitrogen tetroxide dissociates into nitrogen dioxide according to the following equation:

   $$N_2O_{4(g)} \rightleftharpoons 2NO_{2(g)}$$

   When 1 mole of N_2O_4, at an equilibrium pressure of 3.0 atm, is 50.% dissociated, what is the value of K_p?

   Set up an ICE table to find the moles of each substance present at equilibrium.

Answers

1. $K_c = \dfrac{(H_2)(I_2)}{(HI)^2} = \dfrac{(6.3 \times 10^{-4})(1.23 \times 10^{-3})}{(2.50 \times 10^{-2})^2} = 1.25 \times 10^{-3}$

2.

Equation	$2H_2$	S_2	$2H_2S$
Initial moles	1.00	1.00	0.00
Change	−x	$\frac{-x}{2}$	+x
Equilibrium moles	1.00– 0.750	$1-\frac{0.750}{2}$	0.750
Concentration in mol/L at Equilibrium	$\frac{0.250}{1}$ = .250	$\frac{0.625}{1}$ = .625	$\frac{0.750}{1}$ = .750

$$K_c = \frac{[H_2S]^2}{[H_2]^2[S_2]} = \frac{[.750]^2}{[0.250]^2[0.625]} = 14.4$$

3.

ICE Table

	$N_2O_{4(g)} \rightleftarrows$	$2NO_{2(g)}$
Initial	1.0	0
Change	−x	+2x
Equilibrium	1.0 − x	0 + 2x

If we know the gas is 50.% dissociated, then half of it will be converted to NO_2 and x = 0.5. Converting to mole fractions, we find mole fraction of $N_2O_4 = \frac{0.5}{0.5+1.0} = 0.333$, mole fraction of $NO_2 = \frac{1.0}{0.5+1.0} = 0.666$. Then, partial pressure of N_2O_4 = (0.333)(3.0 atm) = 1.0 atm and partial pressure of NO_2 = (0.666)(3.0 atm) = 2.0 atm.

$$Kp = \frac{(2.0)^2}{(1.0)^1} = 4$$

Chapter 24

Le Chatelier's Principle

I. Le Chatelier's Principle

A. Le Chatelier's Principle states that a change in the factors that govern the chemical equilibrium of a system will cause the system to respond in a manner that counteracts the change, i.e., to bring whatever has been changed by the stress (Q—the reaction quotient, or K—the equilibrium constant) back into equality with one another.

1. Chemical equilibrium can be disrupted in different ways.
   - i. Change in temperature
   - ii. Change in concentration of reactant or product
   - iii. Change in volume or pressure (where gases are involved)

   Note: Catalysts do not affect the equilibrium position, only the speed at which it is achieved.

2. Effect of temperature (*will* affect the value of K)

Test Tip

> **Only** ***a temperature change will result in a change of the value of K. This is a common "trick" question. Other changes may cause there to be more reactants or products, i.e., they will affect Q, but if temperature is constant, then K is constant. The equilibrium will always shift in order to bring Q and K back into equality with one another.***

i. Must know whether the reaction is exothermic or endothermic

- Example—Exothermic:

$$O_3(g) + O(g) \rightleftharpoons 2O_2(g) \qquad \Delta H = -318 \text{ kJ}$$

The equation can be rewritten to show that energy is a "product":

$$O_3(g) + O(g) \rightleftharpoons 2O_2(g) + \text{Energy}$$

- If energy is "added" to the system, the result would favor the formation of the reactants or the endothermic reaction.
- If energy is "taken away" from the system, the result would favor the formation of the products or the exothermic reaction.
- Example—Endothermic:

$$2HCl(g) \rightleftharpoons H_2(g) + Cl_2(g) \qquad \Delta H = +185 \text{ kJ}$$

The equation can be rewritten to show that energy is a "reactant":

$$\text{Energy} + 2HCl(g) \rightleftharpoons H_2(g) + Cl_2(g)$$

- If energy is "added" to the system, the result would favor the formation of the products, or the endothermic reaction.
- If energy is "taken away" from the system, the result would favor the formation of the reactants, or the exothermic reaction.

3. Effect of concentration change (*will not* affect the value of K)

Example—$N_2(g) + 3H_2(g) \rightleftharpoons 2NH_3(g)$

- If the concentration of the reactants is decreased, the result would favor the formation of the reactants (reaction shifts backward).

- If the concentration of the reactants is increased, the result would favor the formation of the products (reaction shifts forward).
- If the concentration of the product is decreased, the result would favor the formation of the products (reaction shifts forward).
- If the concentration of the product is increased, the result would favor the formation of the reactants (reaction shifts backward).

4. Effect of volume or pressure change (will *not* affect the value of K)

   i. Must count the number of moles of gas reactants and gas products to accurately predict the direction.

      Example—$2NO_2(g) \rightleftharpoons N_2O_4(g)$

      2 moles of gas reactant, 1 mole of gas product

      - If the volume increases (pressure decreases), the system will shift to the side with the greatest number of gas moles (reactants in the example above) in order to increase the pressure (i.e., shifts to oppose the change).
      - If the volume decreases (pressure increases), the system will shift to the side with the least number of gas moles (products in the example above) in order to decrease the pressure (i.e., shifts to oppose the change).

B. Observing Shifts in Equilibrium

1. Color change—in reactions where reactants and products have different colors, the color of the equilibrium mixture can give clues to the relative proportions of reactants and products.
2. Change in pH—in reactions with H^+ or OH^-, shifts can be detected by monitoring the pH.

Test Tip

Look out for questions where the addition of a species that is not part of the reaction, but does affect the reaction, are added. For example, in the reaction below, H^+ ions are not part of the equilibrium, but when added they will react with hydroxide ions, remove them, and according to Le Chatelier's principle has the effect of shifting the reaction to the product side.

$$Ba(OH)_2(s) \rightleftharpoons Ba^{2+}(aq) + 2OH^-(aq)$$

C. Manipulating Conditions to Optimize Yield

1. Consider the manufacture of ammonia in the Haber Process.

$$N_2(g) + 3H_2(g) \rightleftharpoons 2NH_3(g) \qquad \Delta H = -92\ kJmol^{-1}$$

i. Le Chatelier's principle predicts high pressure will produce more products, but high pressure can be expensive to achieve.

ii. Le Chatelier's principle predicts low temperature will produce more products, but low temperature can mean slow reactions.

iii. Compromise must be reached in order to achieve decent yields in a cost-effective and timely manner.

Practice Question

1. Based on the following reaction, predict whether products or reactants are favored when the stresses listed are applied to the equilibrium system.

$$aA(g) + bB(g) \rightleftharpoons cC(g) \qquad \Delta H = +\ \text{(endothermic)}$$

(a) Decrease in temperature
(b) Increase in pressure
(c) Decrease in volume (same as increase in pressure)
(d) Add more reactant, A
(e) Add a catalyst

Answers

1. (a) Favor reactants

   (b) Favor products

   (c) Favor products

   (d) Favor products

   (e) No change

Chapter 25

Acids, Bases, and Solubility Equilibrium

I. Acids and Bases

A. Strength of Acids and Bases

1. Strong acids (including HCl, HBr, HI, $HClO_4$, H_2SO_4, and HNO_3) and strong bases (including group 1 and most group 2 hydroxides) exhibit complete ionization (dissociation) in water. The group 2 hydroxides that are sparingly soluble in water will still have all of whatever *does* dissolve in the ionized form.
2. For a strong acid (e.g., HCl) and a strong base (e.g., NaOH), the following reactions go to completion (ionization is considered to be 100% with no reverse reaction) to produce large numbers of $H_3O^+_{(aq)}$ and $OH^-_{(aq)}$, respectively.

   $$HCl_{(aq)} + H_2O_{(l)} \rightarrow H_3O^+_{(aq)} + Cl^-_{(aq)}$$

   $$NaOH_{(aq)} \rightarrow Na^+_{(aq)} + OH^-_{(aq)}$$

   The only circumstances where a strong acid would not be 100% ionized would be if the acid were *so* concentrated that the number of water molecules present were insufficient to allow the ionization of every acid molecule.
3. For a weak acid (e.g., CH_3COOH) and a weak base (e.g., NH_3), the following reactions do *not* go to completion (ionization is considered to be very small, perhaps 5% or less, with a reverse reaction that sets up an equilibrium), to produce only small numbers of $H_3O^+_{(aq)}$ and $OH^-_{(aq)}$, respectively.

$$CH_3COOH_{(aq)} + H_2O_{(l)} \rightleftharpoons H_3O^+_{(aq)} + CH_3COO^-_{(aq)}$$

$$NH_{3(aq)} + H_2O_{(l)} \rightleftharpoons NH_4^+{}_{(aq)} + OH^-_{(aq)}$$

4. Examples of weak acids include organic (carboxylic) acids such as methanoic acid (HCOOH), ethanoic acid (CH_3COOH), and propanoic acid (C_2H_5COOH). Examples of weak bases include ammonia (NH_3) and organic bases such as methylamine (CH_3NH_2), ethylamine ($C_2H_5NH_2$), and pyridine.

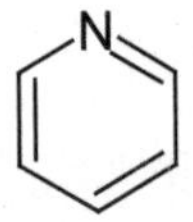

5. Conjugate acid–base pairs are related to one another by a difference of H^+ on either side of the equation. Thus, in the examples above:
   - HCl is the acid and Cl^- is its conjugate base (since HCl is a strong acid, Cl^- is a weak base, and the reaction only goes forward).
   - NH_4^+ is the acid and NH_3 is its conjugate base (since NH_3 is a weak base, NH_4^+ can act as an acid, and the reaction can also go backward).
   - CH_3COOH is the acid and CH_3COO^- is its conjugate base (since CH_3COOH is a weak acid, CH_3COO^- can act as a base, and the reaction can also go backward).

B. pH, pOH, and Kw

1. The pH scale is used to indicate how acidic or basic a substance is, and is defined as:

$$pH = -\log [H_3O^+] \text{ or } pH = -\log [H^+]$$

It is important to note that pH depends upon two, independent factors; first, the degree of ionization (i.e., whether the acid is strong or weak and produces many or few H^+ ions), and, second, the concentration of the solution (i.e., how much water is present). It is possible, for example, for a very concentrated weak acid and a very dilute strong acid to have the same pH, since the concentration of hydronium ions (H_3O^+) could be the same in each case. In one case the concentration of hydronium ions may be primarily due to the concentration of the acid, and in the other primarily due to the strength of the acid.

2. Because bases produce hydroxide ions in solution, the pOH is defined as:

$$\text{pOH} = -\log [OH^-]$$

3. Kw is the ionic product of water. Although pure water is essentially covalent, there is a small amount of self-ionization that occurs between water molecules.

$$H_2O_{(l)} + H_2O_{(l)} \rightleftharpoons H_3O^+_{(aq)} + OH^-_{(aq)}$$

As such, an equilibrium constant, Kw, can be written

$$Kw = [H_3O^+]\,[OH^-]$$

Like all equilibrium constants, Kw is temperature dependent. At 298K, $Kw = 1 \times 10^{-14}$ and $pKw = -\log Kw = 14 = pH + pOH$.

Because pure H_2O will have equal concentrations of $H_3O^+_{(aq)}$ and $OH^-_{(aq)}$, then under these conditions, $[H_3O^+_{(aq)}] = [OH^-_{(aq)}] = \sqrt{1 \times 10^{-14}} = 1 \times 10^{-7}$. Applying $pH = -\log [H_3O^+_{(aq)}]$, we find that the pH of pure water *at 298 K* is 7.

However, since Kw varies with temperature, at other temperatures (i.e., other values of Kw), the pH of water will be a value other than 7, but there will still be equal concentrations of $H_3O^+_{(aq)}$ and $OH^-_{(aq)}$. As such, we define neutrality in terms of there being equal concentrations of $H_3O^+_{(aq)}$ and $OH^-_{(aq)}$ rather than in terms of a particular pH value.

4. For strong acids that completely dissociate because the concentration of the hydronium ions will be the same as the concentration of the acid, simply take the –log of the concentration of the acid to yield the pH. For example, when calculating the pH of a 0.01 M solution of HCl,

$HCl_{(aq)} + H_2O_{(l)} \rightarrow H_3O^+_{(aq)} + Cl^-_{(aq)}$

0.01 0.01

pH = –log (0.01) = 2.00

Test Tip

Because of the math of the pH expression, a change by a factor of 10 in the hydrogen ion concentration will result in a difference of 1 pH unit. For example, –log (0.1) = 1, –log (0.01) = 2, –log (0.001) = 3, etc. Another useful math tip is to know that if the hydrogen ion concentration is expressed in scientific form, e.g., 1×10^{-1}, then the pH will be equal to the absolute value of the power, in this case, pH = 1.

5. Weak acids are not completely dissociated and the equilibrium that exists requires the use of Ka (the acid dissociation constant) to calculate pH. One can set up an ICE table (one that summarizes **I**nitial conditions, **C**hanges in conditions, and **E**quilibrium conditions) to illustrate the reversible reaction of a weak acid, HA, with water.

	$HA_{(aq)}$	+	$H_2O_{(l)}$	$\rightleftharpoons$	$H_3O^+_{(aq)}$	+	$A^-_{(aq)}$
Initial	Y				0		0
Change	– x				+ x		+ x
Equilibrium	Y – x				0 + x		0 + x

i. The acid starts with a concentration of Y and dissociates into ions.

ii. Because it is a weak acid, it does not completely dissociate so it is not possible to go directly to pH = –log [Y] (like one can with strong acids).

iii. However, we do know that because the acid is weak, and there is very little dissociation, *x* will be small and can be considered negligible in the term *Y-x*.

iv. Additionally, we know that the concentrations of $H_3O^+_{(aq)}$ and $A^-_{(aq)}$ will be equal.

v. Considering those facts and knowing that $H_2O_{(l)}$ is a pure liquid (and like all pure liquids and solids it will not appear in the equilibrium expression), one can derive Ka, the acid dissociation constant:

$$Ka = \frac{[H_3O^+][A^-]}{[HA]} \quad \text{or} \quad Ka = \frac{[H_3O^+]^2}{[HA]}$$

Also, pKa = –log Ka.

Test Tip

> *Even within weak acids, there is a hierarchy of strength. Because the strength of an acid is determined by its ability to dissociate and produce H^+ (H_3O^+) ions, and those ions appear in the numerator of the Ka expression, the larger the Ka, the stronger the (weak) acid. However, taking the negative log of the Ka gives pKa, and by means of math we find that the larger the Ka, the smaller the pKa. In summary, the stronger the weak acid, the larger the Ka,* **but** *the larger the Ka, the smaller the pKa.*

vi. Structural factors like the presence of electronegative atoms and bond strength can influence the ability of an acid to lose H^+ and therefore influence the relative strength of a weak acid.

6. For strong bases that completely dissociate, because the concentration of the hydroxide ions will be the same as the concentration of the base, simply take the –log of the concentration of the base to yield the pOH and convert to pH.

i. For example, when calculating the pH of a 0.01 M solution of NaOH:

$$NaOH_{(aq)} \rightarrow Na^{+}_{(aq)} + OH^{-}_{(aq)}$$

0.01   0.01

pOH = –log (0.01) = 2.00,
and at 298 K, pOH + pH = 14,
so pH = 12

7. Weak bases are not completely dissociated and an equilibrium exists which requires the use of Kb (the base dissociation constant) to calculate pH. One can set up an ICE table to illustrate the reversible reaction of a weak base, ammonia, with water.

	$NH_{3(aq)}$	+	$H_2O_{(l)}$	$\rightleftharpoons$	$NH_4^{+}{}_{(aq)}$	+	$OH^{-}_{(aq)}$
Initial	Y				0		0
Change	$-x$				$+x$		$+x$
Equilibrium	$Y-x$				$0+x$		$0+x$

i. The base starts with a concentration of *Y* and dissociates into ions.

ii. Because it is a weak base, it does not completely dissociate, so it is not possible to go directly to pOH = –log [*Y*].

iii. However, we do know that because the base is weak, and there is very little dissociation, *x* will be small and can be considered negligible in the term $Y-x$.

iv. Additionally, we know that the concentrations of $NH_4^{+}{}_{(aq)}$ and $OH^{-}_{(aq)}$ will be equal.

v. Considering those facts, and knowing that $H_2O_{(l)}$ is a pure liquid, and, like all pure liquids and solids, it will not appear in the equilibrium expression, one can derive Kb, the base dissociation constant.

$$Kb = \frac{[OH^-][NH_4^+]}{[NH_3]} \quad \text{or} \quad Kb = \frac{[OH^-]^2}{[NH_3]}$$

Also, pKb = –log Kb.

Test Tip

Other useful relationships for acid–base conjugate pairs include:

Kw = (Ka) (Kb) and pKw = 14 = pKa + pKb

II. Neutralization, Titrations, and Buffers

A. Neutralization

1. The reaction between the $H^+_{(aq)}$ from an acid and $OH^-_{(aq)}$ from a base to produce $H_2O_{(l)}$ is called a neutralization reaction. Commonly, these reactions form a salt in addition to water.
2. A strong acid and a strong base have a neutralization reaction that goes to completion with a $K = 1 \times 10^{14}$, $H_3O^+_{(aq)} + OH^-_{(aq)} \rightarrow 2H_2O_{(l)}$.
3. When a weak acid is neutralized by a strong base, the salt produced will be the conjugate base of the weak acid.
4. When a weak base is neutralized by a strong acid, the salt produced will be the conjugate acid of the weak base.

B. Titrations

1. Titration is the experimental technique for collecting quantitative data about neutralization (and other) reactions.
2. As a base is added to an acid, the acid remains in excess and the pH doesn't change much until the equivalence point is reached. The equivalence point is the point at which the moles of the acid and base have been added in the stoichiometric ratio, neutralization is complete, and at this point there is a rapid change in pH.
3. Since many aqueous solutions of acids and bases are clear, colorless solutions that do not necessarily change color during a reaction, the equivalence point needs to be monitored by an indicator. An indicator is a chemical (often a weak acid itself) that changes color over a narrow pH range. The observable point at which the indicator changes color is called the end point, and an indicator must be chosen that changes color at a pH as close to the equivalence point as possible.
4. For a strong and weak acid of the same pH, more base will be required to neutralize the weak acid, since to have the same pH as the strong acid, the weaker acid must be more concentrated.
5. Typical titration curves (plots that show the variation of pH with added base) are shown below.

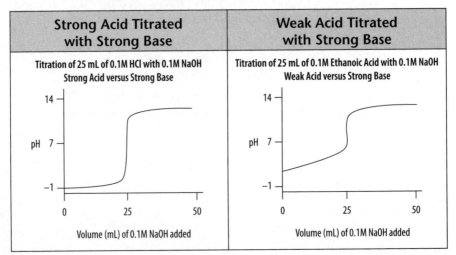

Strong Acid Titrated with Strong Base	Weak Acid Titrated with Strong Base
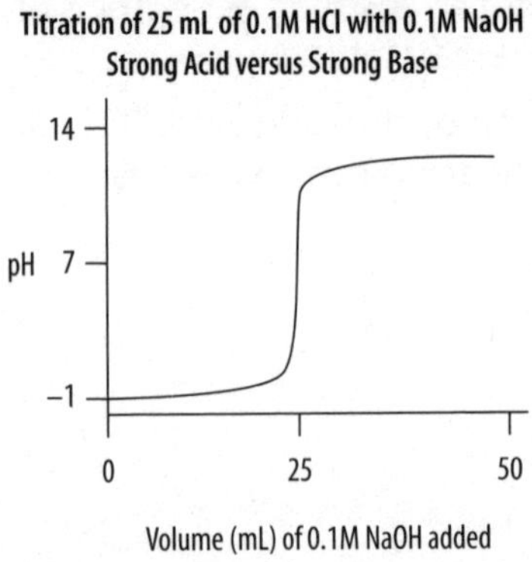	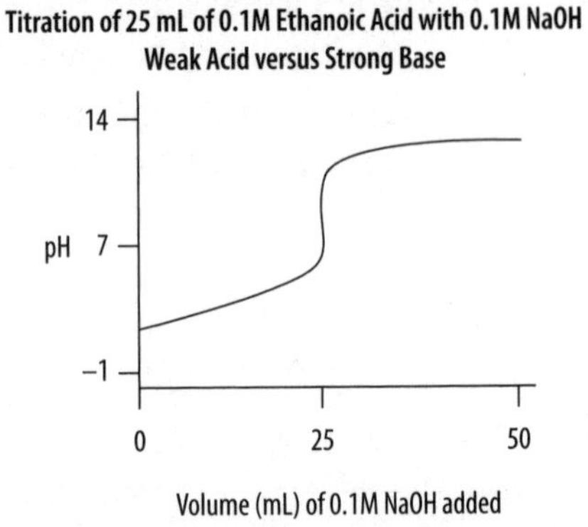

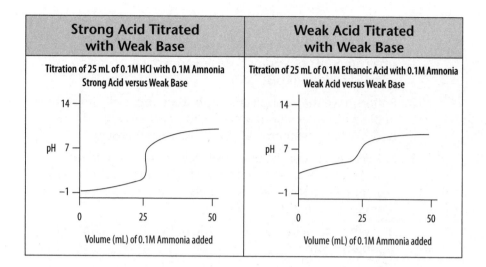

Strong Acid Titrated with Weak Base

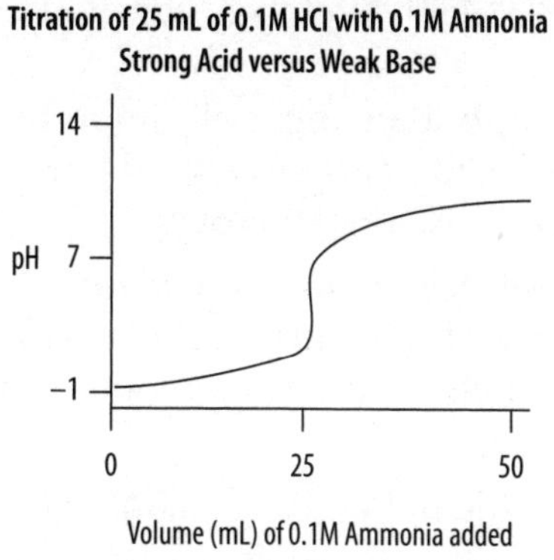

Weak Acid Titrated with Weak Base

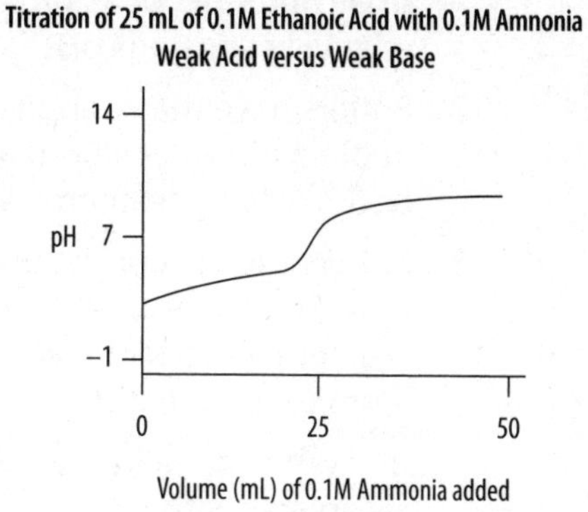

6. The titration curve for a polyprotic acid (one with more than on H^+ ion to donate) will be similar, but will have multiple equivalence points (one for each H^+ ion that it can donate to the base).

Titration of Phosphoric Acid with Base

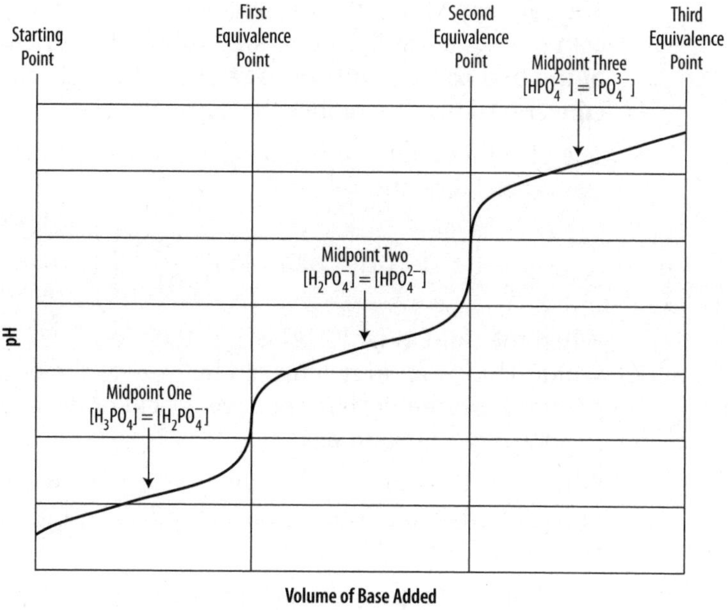

C. Buffers

1. A buffer solution is one that resists change in pH when small amounts of acid or base are added, i.e., it maintains a relatively constant pH.
2. Buffers have many applications, but are especially important in biochemistry (blood, amino acids, and proteins in the body). Many biochemical reactions are pH sensitive.
3. Buffers usually consist of a weak acid and its conjugate base (a salt), or a weak base and its conjugate acid (a salt), e.g., ethanoic acid and sodium ethanoate or ammonia and ammonium chloride.
   i. When an acid is added to the buffer, the conjugate base reacts with it.

$$H^+ + CH_3COO^- \rightarrow CH_3COOH$$

   ii. When a base is added to the buffer, the conjugate acid reacts with it.

$$OH^- + CH_3COOH \rightarrow CH_3COO^- + H_2O$$

4. The capacity of a buffer (its ability to continue to absorb any acid or base added) is dependent upon the concentrations of the two components of the buffer; the higher the concentrations, the higher the capacity.
5. The pH of a buffer can be calculated using the Henderson–Hasselbalch equation.

$$\text{pH} = \text{pKa} + \log\left(\frac{[A^-]}{[HA]}\right)$$

When the ratio of $[A^-]:[HA]$ is 1:1, then log (1) = 0 and pH = pKa. This position is achieved halfway to the equivalence point during the titration of a weak acid or weak base with a strong base or strong acid.

Because of the math of the equation, a measurement of the pH of a buffer solution can give an indication of the relative concentrations of A^- and HA.

   i. When pH > pKa, then $[A^-] > [HA]$

ii. When pH < pKa, then $[A^-] < [HA]$

One can determine which species is present in the higher concentration, e.g., where labile protons might be found.

6. Buffers and titration curves. When a weak acid or a weak base is titrated with a strong base or strong acid, a salt (the conjugate) is produced in solution. Since all the way up to the equivalence point excess acid or base is present, a buffer is created.

III. Solubility and Ksp

A. Ksp—Solubility Product Constant

1. All sodium, potassium, ammonium, and nitrate salts are soluble in water.
2. Some other salts are considered insoluble, but all solubilities can be expressed in terms of equilibrium. For example, a salt, MX, will form equilibrium in water, thus:

$$MX_{(s)} \rightleftharpoons M^+_{(aq)} + X^-_{(aq)}$$

Since it is considered an equilibrium, it can be assigned values for K (known as Ksp) and Q. The extent to which the equilibrium lies on one side or the other determines the relative solubility of the salt. Larger Ksp values suggest greater dissociation into ions and greater solubility. For the reaction above,

$$Ksp = [M^+_{(aq)}]\ [X^-_{(aq)}]$$

In general, for the reaction:

$$A_xB_{y(s)} \rightleftharpoons xA^{y+}_{(aq)} + yB^{x-}_{(aq)}$$

$$Ksp = [A^{y+}]^x\ [B^{x-}]^y$$

For example, AgCl has a Ksp = 1.77×10^{-10}, meaning:

$$Ksp = [Ag^{+}_{(aq)}]\ [Cl^{-}_{(aq)}] = 1.77 \times 10^{-10}, \text{ and}$$

$$[Ag^{+}_{(aq)}] = [Cl^{-}_{(aq)}] = 1.33 \times 10^{-5}\ M$$

which is known as silver chloride's solubility. The larger the individual ion concentrations at equilibrium, the larger the Ksp value and the more soluble the salt. It should also be noted, as with all equilibrium constants, that Ksp values and consequent solubilities are temperature dependent.

- When Q = Ksp, the equilibrium has been established and all ion concentrations are constant.
- When Q > Ksp, the product of the equilibrium ion concentrations is too large and the equilibrium must be established by reducing those concentrations and forming more solid (i.e., precipitation occurs).
- When Q < Ksp, the product of the equilibrium ion concentrations is too small and the equilibrium must be established by increasing those concentrations and forming more ions, i.e., no solid (precipitate) is formed.

3. The relative thermodynamic favorability of the dissolution of a salt is determined by the sign of ΔG, which depends on both enthalpy and entropy. The enthalpy change for dissolution is dependent upon (and a sum of) three independent factors:
   i. The separation of the solute particles from one another (for example, breaking ionic bonds in an ionic substance, an endothermic process)
   ii. The separation of the solvent particles from one another (for example, breaking hydrogen bonds in water, an endothermic process)
   iii. The interaction between the solute particles and the solvent particles (for example, ions becoming hydrated [surrounded by water molecules] when an ionic substance dissolves in water, an exothermic process)

In addition to enthalpy changes, we need to consider entropy changes and, when a solute dissolves, entropy (disorder) increases. It is the cumulative effect of enthalpy *and* entropy factors at a given temperature that determine ΔG and, ultimately, the thermodynamic favorability of any dissolution process.

4. The common ion effect and pH. A salt becomes increasingly *less* soluble in a solution that contains one of its ions. This can be explained in terms of Le Chatelier's principle and is called the common ion effect. For example, some chlorides are less soluble in sea water (that contains chloride ions) than pure water, since the presence of chloride ions will force the equilibrium to the reactant (solid) side;

$$MCl_{2(s)} \rightleftharpoons M^{2+}_{(aq)} + 2Cl^{-}_{(aq)}$$

Similarly, when a salt contains ions that can act as an acid or a base, the solubility will be affected by pH. For example, iron (III) hydroxide will be less soluble in basic solution, since Le Chatelier's principle predicts that the equilibrium will be shifted to the reactant (solid) side.

$$Fe(OH)_{3(s)} \rightleftharpoons Fe^{3+}_{(aq)} + 3OH^{-}_{(aq)}$$

Also, solid calcium carbonate is an example of a salt that is almost entirely insoluble in water, but in acid rain (i.e., at low pH's) it will dissolve. As such, its solubility is highly pH dependent.

Practice Questions

1. A pure 1.72 g sample of a weak acid, HNO_2 (nitrous acid), is dissolved in enough water to make 250 mL of solution.
   (a) Calculate the concentration of HNO_2.
   (b) Nitrous acid reacts with water according to the equation below:

   $$HNO_2(aq) + H_2O(l) \rightleftharpoons H_3O^+(aq) + NO_2^-(aq),\ K_a = 6.0 \times 10^{-4}$$

   Write the equilibrium-constant expression for the reaction of $HNO_2(aq)$ and water.
   (c) Determine the pH of the solution in (b).

2. A weak monoprotic acid, HA(aq), dissociates in water as represented below:

   $$HA(aq) + H_2O(l) \rightleftharpoons H_3O^+(aq) + A^-(aq)$$

   $$K_a = 7.2 \times 10^{-4}$$

   (a) Write the equilibrium constant-expression for the dissociation of HA(aq) in water.
   (b) Calculate the molar concentration of H_3O^+ in 0.25M HA(aq) solution.

3. The solubility of AgBr is 5.74×10^{-7} mol L^{-1} at 25°C. Calculate the K_{sp} of AgBr.

4. Calcium fluoride will dissolve in water. If the concentration of Ca^{2+} is 1.34×10^{-2} M, find the K_{sp} for calcium fluoride.

5. If the K_{sp} for BaF_2 is 1.7×10^{-6}, determine the solubility of the salt.

6. Consider the two organic acids shown below.

Ethanoic Acid	Chloroethanoic Acid
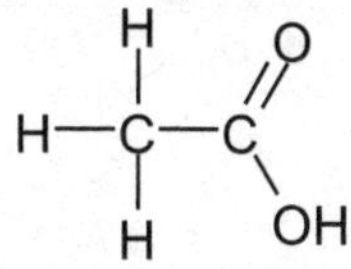	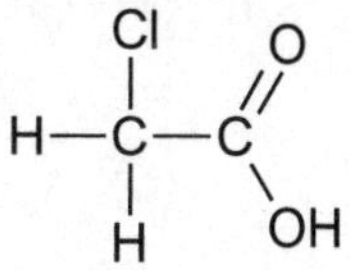
pKa = 4.76	pKa = 2.86

(a) Identify the stronger acid and account for the differences in the pKa's of the two acids.

(b) When ethanoic acid (a weak acid) is titrated with a strong base (NaOH), the equivalence point is found to be at a pH of approximately 9.

(i) Using the table below, choose a suitable indicator for the titration of ethanoic acid with sodium hydroxide. Justify your choice.

Indicator	pKa
Methyl Orange	3.7
Methyl Red	5.1
Bromothymol Blue	7.0
Phenolphthalein	9.3

(ii) It is found that the pH of the acid before any base has been added is 2.9, and that the volume of NaOH required to reach the end point is 20 mL. Sketch the titration curve that you would expect to be generated during this titration, using "pH" as the *y*-axis and "Volume of NaOH added in mL" as the *x*-axis, when 30 mL of base is added to the ethanoic acid.

(iii) Distinguish between the terms "end point" and "equivalence point." What is their relationship to one another?

Answers

1. (a) $\frac{1.72\text{ g HNO}_2}{2} \times \frac{1\text{ mol HNO}_2}{47.018\text{ g HNO}_2} = \frac{0.0366\text{ mol HNO}_2}{0.250\text{ L}} = 0.146\text{ M HNO}_2$

(b) $K_a = \frac{[NO_2^-][H_3O^+]}{[HNO_2]}$

(c)

Equation	HNO_2	H_3O^+	NO_2^-
Initial concentration	0.146	0	0
Change	$-x$	$+x$	$+x$
Equilibrium concentration	$0.146 - x$	$+x$	$+x$

$$K_a = \frac{[x][x]}{0.146 - x} = 6.0 \times 10^{-4}$$

Since the acid is weak and the degree of dissociation is negligible, one can assume $x < 0.146$, so the denominator can be expressed simply as 0.146.

$x^2 = (6.0 \times 10^{-4})(0.146)$

$x = 0.0094 = H_3O^+$

$pH = -log[H_3O^+] = -log\ 0.0094 = 2.03$

2. (a) $K_a = \frac{[A^-][H_3O^+]}{HA}$

(b) $K_a = \frac{[x][x]}{0.25} = 7.2 \times 10^{-4}$

$x^2 = 1.8 \times 10^{-4}$

$x = 0.013 = H_3O^+$

3. $AgBr(s) \rightleftharpoons Ag^+(aq) + Br^-(aq)$

   $K_{sp} = [Ag^+][Br^-]$

   $K_{sp} = [5.74 \times 10^{-7}]\,[5.74 \times 10^{-7}] = 3.29 \times 10^{-13}$

4. $CaF_2(s) \rightleftharpoons Ca^{2+}(aq) + 2F^-(aq)$

   $K_{sp} = [Ca^{2+}][F^-]^2$

   $K_{sp} = [1.34 \times 10^{-2}]\,[(2)1.34 \times 10^{-2}]^2 = 9.62 \times 10^{-6}$

5. $BaF_2(s) \rightleftharpoons Ba^{2+}(aq) + 2F^-(aq)$

   $K_{sp} = [s][2s]^2$

   $1.7 \times 10^{-6} = 4s^3$

   $s = 7.5 \times 10^{-3}M\ BaF_2$

6. (a) Chloroethanoic acid is the stronger acid and hence has a smaller pKa.

   The chlorine atom is electronegative and draws electron density away from the –COOH group. This weakens the O–H bond within –COOH, meaning that H^+ ions are released more readily. As a result, chloroethanoic acid is a stronger acid than ethanoic acid, since ethanoic acid lacks Cl atoms, therefore, experiences no such electron withdrawing effect, and no such weakening of the O–H bond occurs.

   In addition, by withdrawing electrons from the $-COO^-$ part of the carboxylic acid ion formed when the acid gives up hydrogen ions, the electronegative chlorine atom has the effect of dispersing the build up of charge around $-COO^-$, and hence the ion is stabilized. This means that the ion is more likely to exist as an entity, is less attractive to the H^+ ions that have been released, and that the acid is more likely to remain dissociated.

(b)(i) Phenolphthalein. An indicator should be chosen that has a pKa as close to the pH of the equivalence point as possible.

(ii)

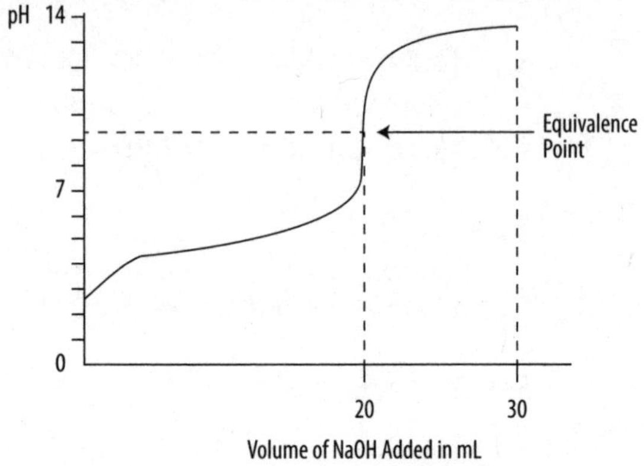

(iii) The end point is the point in a titration when some observable change takes place, usually the point at which an indicator changes color. The equivalence point is the point at which the titrant has completely reacted with the analyte in the stoichiometric amounts, in this case when the base has completely neutralized the acid. When performing a titration, it is desirable for the end point to occur as close to the equivalence point as possible.

Chapter 26

Equilibrium and Gibbs Free Energy

I. The Relationship Between ΔG° and K

A. Mathematical Relationships

1. Standard Gibbs Free Energy Change and K are related by the equation below and is the change in free energy when all substances in the reaction are in their standard states under standard conditions of 1 M solutions and 1 bar/1atm pressure.

   $\Delta G^\circ = -RT \ln K$

   R = universal gas constant = 8.314 J mol^{-1} K^{-1}

   T = temperature in Kelvin

   K = equilibrium constant

   The equation can be re-arranged to give:

   $K = e^{-\Delta G^\circ/RT}$

2. If K is to be close to 1, i.e., significant amounts of *both* products and reactants are present at equilibrium, then ΔG° must be approximately equal to RT (the thermal energy), i.e., 2.4 kJ/mol at room temperature, 298K.

   Under these circumstances, products and reactants are approximately equally favored at equilibrium.

3. If ΔG° is much more positive than 2.4 kJ/mol, then K is very small and there are very few products and a large proportion of reactants at equilibrium, i.e., when $\Delta G^\circ > 0$, reactants are favored and $K < 1$.

4. If $\Delta G°$ is much more negative than 2.4 kJ/mol, then K is very large and there are very few reactants and a large proportion of products at equilibrium, i.e., when $\Delta G° < 0$, products are favored and $K > 1$.
5. When $\Delta G°$ is positive, the reaction is called *endergonic* and $K < 1$ (favors reactants).
6. When $\Delta G°$ is negative, the reaction is called *exergonic* and $K > 1$ (favors products).

PART VIII

OVERARCHING THEMES

Chapter 27

Laboratory Work in AP Chemistry

I. Lab Program

A. Overview

1. If your course is one that has been audited by the College Board (one that allows your institution to put "AP" on your transcript), then you should have experienced the following features:

   i. Laboratory takes 25% of your instructional time.

   ii. A minimum of 6 labs should be inquiry labs ("inquiry" has a wide-reaching meaning and can take many forms).

   iii. A total of 16 labs should have been conducted.

   If your course has been audited by the College Board as an "alternative approach," then your lab program may have included demos, simulations, and other virtual experiences that have replaced some of the traditional lab work.

II. Science Practices

The College Board lists seven *science practices* that you should have been exposed to during your AP course. These are largely common-sense things that will automatically form part of any competent AP experience, but they have been formalized and elaborated upon in the new curriculum. Let's examine each science practice and list the things that you should be able to do in relation to each one.

1. *The student can use representations and models to communicate scientific phenomena and solve scientific problems.*

   Can you visualize, draw, and use models of particles (atoms, molecules, ions) to explain macroscopic changes that you observe in the lab? There is a very large emphasis in the new curriculum on the visualization and modeling of the microscopic via macroscopic, particulate diagrams. Expect to see a lot of diagrams on the exam, with shapes used to represent particles. A couple of examples are a diagram showing water molecules surrounding ions when an ionic solid is dissolved into aqueous solution, or using VSEPR to predict the shape and polarity of molecules.

2. *The student can use mathematics appropriately.*

   Can you use quantitative data that you collect during experiments to calculate other related values? Can you use significant figures correctly and can you appreciate that accuracy and precision are important considerations when considering which equipment (e.g., glassware) to use in any given situation? Examples include calculating an empirical formula from experimental data or making an estimation about a numerical value in a multiple-choice question. You should appreciate that certain combinations of signs of enthalpy and entropy have consequences for the sign of Gibbs Free Energy via the equation that links them together.

3. *The student can engage in scientific questioning to extend thinking or to guide investigations within the context of the AP course.*

   Can you formulate a question or hypothesis (and subsequently an experimental procedure; see science practice 4 below) that can be investigated in the lab? For example, can you suggest the variables to consider in order to investigate the factors that determine the rate of a chemical reaction?

4. *The student can plan and implement data collection strategies in relation to a particular scientific question. (Note: Data can be collected from many different sources, e.g., investigations, scientific observations, the findings of others, historic reconstruction, and/or archived data.)*

Having identified variables to investigate from the questions that you have posed in science practice 3, can you develop an experimental procedure to collect that data? For example, can you choose and then justify your choice of which variables to control, which to change, and what to measure in the investigation of reaction rate in a kinetics experiment? Can you then design an experiment around those choices in order to determine a rate or order of a chemical reaction?

5. *The student can perform data analysis and evaluation of evidence.*

   Can you identify trends or patterns in data, including the identification of data that may be classified as outliers? Can you utilize tools such as graphs and data in charts or tables, or perform calculations with the data to help you evaluate the results of an experimental procedure? For example, having collected the data in an experiment that you may have designed in science practice 4, can you analyze it to observe a pattern that leads to a conclusion, or manipulate it to find a relationship between different variables?

6. *The student can work with scientific explanations and theories.*

   Can you link the data that you collect in an experiment to an accepted theory or explanation? Does your data match what we already know to be the accepted explanation? If not, what are the reasons for the discrepancy and can you identify possible sources of error? For example, can you link a theory such as Le Chatelier's principle to predict how a chemical reaction will shift when a stress is placed upon it during a laboratory situation?

7. *The student is able to connect and relate knowledge across various scales, concepts, and representations in and across domains.*

   Can the models and data that you collect be expanded into other areas or situations? Can you see how the data that you collected in an isolated experiment in the lab can be applied in other situations outside of the lab work that you have conducted? For example, can you relate the macroscopic, observable properties of substances to the microscopic interpretation of their structures, such as linking the relative strength of intermolecular forces between the molecules of covalently bonded liquids to their measurable boiling points?

III. The College Board Lab Manual

The College Board lab manual offers 16 laboratory investigations as examples of the laboratory experience that you may have had during your course. These are not "required" labs as such. Rather, they are examples of labs you may have experienced.

Each of the labs in the manual is offered in an inquiry manner, which may or may not be the way that the lab was presented to you. "Inquiry" can take on many different guises, but typical characteristics are a question being posed and then a student developing a procedure based upon their prior knowledge that will allow them to collect and analyze data in order to answer the original question. The level of support and guidance offered by the teacher is what makes the distinction between the different levels of guided inquiry and completely open-ended inquiry.

Below are questions that could be the stimulus for 16 lab situations and the thought processes that will help you to answer lab situation questions that might appear on the AP Chemistry exam.

1. **How does the concentration of a colored solution relate to absorbance?**

   What is this about?

   The Beer-Lambert law and the relationship between the concentration of a colored solution and the amount of light that it absorbs

   What do I need to know?

   $A = a\,b\,c$;

   There is a linear relationship between A and c;

   The optimum wavelength to use in a colorimeter or spectrophotometer is that of the complementary color to the color of the solution (i.e., the color that gives the highest absorbance);

The procedure for diluting solutions including appropriate glassware (buret, pipet, volumetric flask);

This procedure is only useful for colored solutions, often those of transition metals.

What might I be asked?

The procedure and calculations associated with dilution, including the use of appropriate glassware and the order of mixing;

To plot or interpret a graph of absorbance against concentration;

To select an appropriate wavelength of incident light.

Sample Question

Why are solutions of sodium chloride and magnesium nitrate unsuitable for analysis via a colorimeter in a Beer-Lambert law experiment?

Answer

Both solutions are colorless and, as a result, their absorbance will not vary with concentration of solution.

2. What is the composition of an alloy?

What is this about?

The analysis of an alloy using the Beer-Lambert law.

What do I need to know?

Alloys are mixtures of metals;

That an alloy can be dissolved in an acid to produce a solution of metals ions and if the resulting solution is colored, then it can be analyzed using the Beer-Lambert law as outlined in experiment 1, above.

What might I be asked?

Same as in 1, above.

Sample Question

Brass is an alloy that contains copper metal. Copper(II) ions impart a blue color to an aqueous solution that includes them, and, as such, a solution containing them can be used to determine concentration in a Beer-Lambert experiment. A common method of converting copper metal to copper(II) ions is to react copper metal with concentrated nitric acid. Write an equation to show this reaction and identify the type of reaction taking place.

Answer

$$Cu_{(s)} + 4HNO_{3(aq)} \rightarrow Cu(NO_3)_{2(aq)} + 2NO_{2(g)} + 2H_2O_{(l)}$$

REDOX

3. **How can we determine the concentration of an aqueous ion in solution?**

What is this about?

The stoichiometry, and prediction of, precipitation reactions.

What do I need to know?

Solubility rules;

The theory of double displacement precipitation reactions, and how to write a net ionic equation;

Simple stoichiometry of solutions;

Procedures for simple gravimetric analysis including adding excess reactant to ensure all target ions are precipitated, correct filtering procedure, and the need for washing, drying, and careful weighing of precipitates.

What might I be asked?

To select an appropriate ionic solution to precipitate the ion that is to be analyzed;

To write full, ionic, and net ionic equations;

The procedure for precipitation;

Stoichiometric calculations relating to concentration, volume, mass, moles, and percentages.

Sample Question

Write the net ionic equation for the reaction of barium nitrate solution with sodium sulfate solution.

Answer

$$Ba^{2+}{}_{(aq)} + SO_4{}^{2-}{}_{(aq)} \rightarrow BaSO_{4(s)}$$

4. **What is the concentration of an acid?**

What is this about?

The titration of an acid and a base.

What do I need to know?

Neutralization reactions, i.e., acid + base → salt + water;

Titration procedure, including washing and filling burets and pipets, and selection of indicators;

How to prepare a standard solution;

Titration curves including pH and buffer implications.

What might I be asked?

To write a neutralization equation;

To detail the procedure of a titration (rinsing and filling burets and pipets including filling the tip of a buret; using a pipet to accurately dispense the analyte, and the buret to accurately dispense the titrant; the use of an Erlenmeyer flask as the reaction vessel; the addition of a few drops of a suitable indicator (see Chapter 25); the potential for the use of a white tile to help observe the end point);

To perform calculations relating to concentration, volume, and pH (Henderson-Hasselbalch).

Sample Question

In the titration of a strong acid using a strong base as the titrant, a student uses an indicator that changes the color at a pH of approximately 2. What effect does that have on the volume of titrant added to the flask? (Reference the terms "end-point" and "equivalence point" in your answer.)

Answer

Too little titrant will be added. The end-point will occur long before the equivalence point of approximately 7.

5. **What are the components of the mixture that makes up a food dye?**

What is this about?

Chromatography as a separation technique.

What do I need to know?

Chromatography involves a moving phase and a stationary phase, and separation depends on the components affinity for one or the other;

How to calculate an R_f value.

What might I be asked?

To interpret a chromatogram to analyze a mixture;

To calculate a R_f value;

To outline the procedure for a simple chromatography experiment;

To choose an appropriate solvent to separate a mixture based on polarities.

Sample Question

When comparing the R_f values of two components of a mixture on a single chromatogram, what does a large R_f value for one of the components tell us?

Answer

The larger the R_f value, the further the component traveled with the solvent. Subsequently, it can be deduced that the component with the larger R_f value has either a higher affinity for the moving phase or a lower affinity for the stationary phase.

6. **How can we relate bonding to properties?**

What is this about?

The influence that bonding has on observable and measurable properties.

What do I need to know?

The general properties and characteristics associated with ionic, covalent, and metallic materials.

What might I be asked?

To predict or explain behavior of materials in relation to properties such as electrical and thermal conductivity, solubility in polar and nonpolar solvents, hardness, melting point, boiling point, etc.

Sample Question

In an experiment, it is found that a particular white, crystalline solid does not conduct electricity when it is a solid, but does conduct electricity when it is in solution or molten. What does this observation suggest about the bonding present in the solid?

Answer

It is predominantly ionic since in an ionic solid the ions cannot freely move and cannot conduct electricity. However, when in solution or molten, the ions can move freely and they can conduct.

7. How efficient are chemical processes?

What is this about?

Stoichiometry, percentage yield, and atom economy.

What do I need to know?

Solubility rules, precipitation reactions, and how to write a net ionic equation;

Simple stoichiometry of solutions;

Procedures for simple gravimetric analysis including adding excess reactant to ensure all target ions are precipitated, correct filtering procedure, and the need for washing, drying, and careful weighing of precipitates;

Percentage yield calculations;

Atom economy calculations;

$$\text{Percent Atom Economy} = \frac{\text{Mass of desired product}}{\text{Mass of all products}} \times 100;$$

How to perform a calculation relating to decomposition of hydrated salts.

What might I be asked?

To find x in the formula $MgSO_4 \cdot xH_2O$;

Stoichiometric calculations relating to concentration, volume, mass, moles, and percentage yield, and percent atom economy.

Sample Question

What is meant by *"heating to constant mass"* in an experiment designed to analyze the number of moles of water present in a hydrated salt?

Answer

Heating the salt until all of the water of crystallization has been driven off, so only the anhydrous salt remains and the mass of the salt ceases to change.

8. **How can we quantitatively analyze substances using a reduction-oxidation reaction?**

What is this about?

REDOX reactions and their analysis via titrations.

What do I need to know?

A definition and understanding of oxidation and reduction;

How to write and combine half-reactions;

Common oxidizing and reducing agents and their half-reactions;

A procedure for titration.

What might I be asked?

To write half-reactions;

To combine half-reactions;

To perform calculations relating to titration data (reacting ratios, concentration, and volume, etc.)

Sample Question

When using potassium manganate(VII) (potassium permanganate) as the titrant in the titration of a solution of iron(II) ions, what is observed at the end-point?

Answer

A permanent light pink color caused by the final drop of manganate(VII) (potassium permanganate) solution being in excess after the iron(II) ions have been exhausted.

9. Separation of Mixtures

What is this about?

Separation techniques based upon solubility of components;

Intermolecular and intra forces.

What do I need to know?

Ionic salts will tend to dissolve in polar solvents such as water;

Large, covalent molecules (often organic in nature) will tend to dissolve in nonpolar (often organic) solvents;

Simple filtration techniques;

Simple liquid-liquid separation techniques including knowledge of a separating funnel and its use;

Basic, gravimetric analysis including filtering, drying, and massing.

What might I be asked?

To interpret mass data collected via gravimetric analysis;

To predict in which layer (aqueous or organic) various components of a mixture might dissolve.

Sample Question

During gravimetric analysis, why is it important to fully dry, but not excessively heat, solids that are collected by filtration of an aqueous solution?

Answer

If solids are not completely dry, they will include water, and massing them will lead to anomalously high values being recorded. Excessive heating could cause an unstable solid to decompose, causing a mass to be recorded that is too small.

10. How do factors like particle size, concentration, and temperature affect the speed of a chemical reaction?

What is this about?

Kinetics and which factors affect the speed of a reaction.

What do I need to know?

How factors such as particle size, temperature, concentration, and catalysts affect the rate of a reaction;

How those factors are explained at the microscopic level based on chemistry concepts (collision theory, activation energy, etc.);

Dilution techniques and procedures.

What might I be asked?

To design an experiment with controls or to investigate the speed of a chemical reaction (including how to measure the rate of reaction);

To interpret data that is generated in such an experiment including graphical representations.

Sample Question

Explain why using powdered calcium carbonate, as opposed to large chips of calcium carbonate, makes the reaction between $CaCO_{3(s)}$ and aqueous hydrochloric acid proceed at an increased rate.

Answer

The powdered calcium carbonate has a greater surface area than the large chips, and, as a result, far more collisions can take place between the acid and the solid. An increased number of collisions means a faster rate.

11. How can we deduce a rate law experimentally?

What is this about?

Using initial rate and concentration data to determine a reaction rate law.

What do I need to know?

Rate laws can be determined by the initial rate method;

The relationship between changes of rate and orders of reaction;

The shape and interpretation of graphs as they relate to the zeroth, first, and second order reactions.

What might I be asked?

To interpret concentration data generated in, for example, a Beer-Lambert law experiment, and to use it to determine a rate law;

To interpret or plot a graph to determine a rate law.

Sample Question

Having collected the data in a kinetics experiment, a student plots the reciprocal of the concentration of reactant A (*y*-axis) against time on the (*x*-axis), and finds that a straight line is *not* produced. What does this information *alone* tell the student about the order of reaction with respect to the concentration of reactant A?

Answer

The order of reaction with respect to reactant A is *not* second order. If it *were* second order, then a straight line would result in this plot.

12. What is the energy change during the process of dissolving a salt in water?

What is this about?

Enthalpy of reaction and calorimetry.

What do I need to know?

The application of $q = m\ c\ \Delta T$;

The relationship between q and enthalpies measured in kJ/mol;

A procedure for calorimetry;

The energy changes associated with dissolving an ionic solid;

Certain glassware offers certain degrees of precision.

What might I be asked?

To design an experiment (including selection of concentrations and volumes of solutions) in order to collect temperature change data and convert that data to energy measurements;

To apply $q = m\, c\, \Delta T$;

To select glassware based upon the accuracy and precision required.

Sample Question

The standard enthalpy of neutralization for a strong acid with a strong base is approximately –57 $kJmol^{-1}$. When a neutralization reaction is repeated using a very weak acid such as HCN and a strong base, the enthalpy of neutralization is found to be only approximately –12 $kJmol^{-1}$. Explain the difference.

Answer

The weak acid HCN will only be partially ionized and, as such, some energy will be required to dissociate it before complete neutralization can take place. This results in a less exothermic reaction.

13. What causes the equilibrium position to shift?

What is this about?

Le Chatelier's principle.

What do I need to know?

Le Chatelier's principle and predicting how changes in conditions (stresses) affect the position of an equilibrium;

Macroscopic observations (such as color changes, pH changes, etc.) can be used to determine the shift in an equilibrium position.

What might I be asked?

To predict or explain shifts in equilibrium when changes in temperature, concentration, pressure, or catalysts are applied to a reaction and to be able to explain each in terms of Le Chatelier's principle;

To interpret particulate diagrams that show relative numbers of species as a function of equilibrium position.

Sample Question

Oxygen in the human body forms an equilibrium mixture with hemoglobin (Hb) represented by the equation below:

$$Hb_{(aq)} + O_{2(g)} \rightleftharpoons HbO_{2(aq)} \qquad K_{eq} = x$$

Carbon monoxide gas creates a similar equilibrium with hemoglobin, thus:

$$Hb_{(aq)} + CO_{(g)} \rightleftharpoons HbCO_{(aq)} \qquad K_{eq} = 200x$$

Calculate the equilibrium constant for the reaction

$$HbO_{2(aq)} + CO_{(g)} \rightleftharpoons HbCO_{(aq)} + O_{2(g)}$$

Answer

Since the final equation is the first equation reversed, and added to the second equation, we can take the reciprocal of the first equilibrium constant and multiply it by the second equilibrium constant, to achieve the equilibrium constant for the final reaction.

$$\left(\frac{1}{x}\right)(200x) = 200$$

14. What do titration curves tell us about reactions?

What is this about?

Titration curves of weak and strong acids and bases.

What do I need to know?

The shapes of titration curves for any combination of weak and strong acids;

What the dominant species are in solution at all points on the curve (including when titrating weak acids and weak bases, halfway to the equivalence point);

That when dealing with weak acids and weak bases in titrations with strong bases and strong acids, respectively, that buffer solutions are produced;

The Henderson-Hasselbalch equation;

The difference between *equivalence point* and *end point.*

What might I be asked?

To sketch and interpret titration curves;

To relate the titration curve to pH and pKa values;

To identify dominant species at any point in a titration;

To be able to draw particulate diagrams to demonstrate your knowledge of dominant species;

To perform titration calculations (concentrations, volumes, molar ratio, etc.).

Sample Question

When a weak acid is titrated with a strong base, what can be said of the composition and the pH of solution in the Erlenmeyer flask when the titration is halfway to the equivalence point?

Answer

It will have equal concentrations of acid and conjugate base, and the pH will equal the pKa of the weak acid.

15. What makes a good buffer?

What is this about?

Buffers

What do I need to know?

What a buffer is;

How a buffer works and how to write equations to show the buffering action;

The factors that affect a buffer's capacity;

The factors that affect a buffer's pH (i.e., the application of the Henderson-Hasselbalch equation);

Where solutions are buffered in relation to titration curves;

The definition of a polyprotic acid;

pH = pKa halfway to the equivalence point.

What might I be asked?

To interpret or sketch a titration curve in terms of the buffering action possible when a weak acid or weak base is titrated with a strong base or strong acid;

To perform calculations using the Henderson-Hasselbalch equation;

To identify the pH at the equivalence points of various titration curves.

Sample Question

Which of the following pairs will make the best buffer solution? Explain your answer.

$NaOH$ and NH_3

CH_3COOH and CH_3COONa

HCl and $NaCl$

CH_3COOH and $NaCl$

Answer

CH_3COOH and CH_3COONa

This combination is that of a weak acid and its conjugate base.

16. What affects the pH and capacity of a buffer?

What is this about?

The capacity and pH of buffers.

What do I need to know?

How to interpret the Henderson-Hasselbalch equation in terms of buffer capacity (the concentration of each component) and the desired pH (ratio of each component).

What might I be asked?

How to perform calculations with the Henderson-Hasselbalch equation and to suggest good combinations of weak acid/base and conjugates, both in terms of concentrations (capacity) and ratio (pH) to achieve desired buffering activity.

Sample Question

Why might it be desirable to have a buffer with approximately equal amounts of weak base and conjugate acid?

Answer

The buffer would have approximately equal capacity to absorb any acid or base that was added to it.

IV. General Laboratory Procedures and Practice

A. Lab Equipment

It is not uncommon for AP questions to ask you to match a piece of glassware or lab equipment to a particular function or purpose. You should become familiar with names and appearance of equipment in the following list, along with their function and role in the lab. Also note that different glassware has different degrees of precision and accuracy.

Name of Equipment	Picture	Function
Beaker	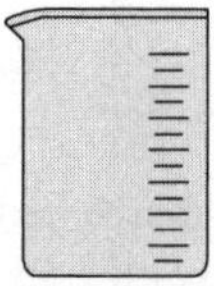	Glass used to hold and heat solutions. Not used for measuring.
Büchner Funnel	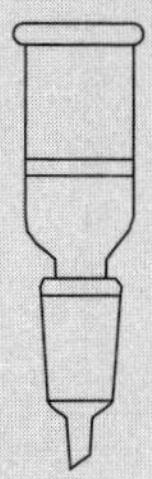	Used for suction filtration along with a filtering flask.
Bunsen Burner	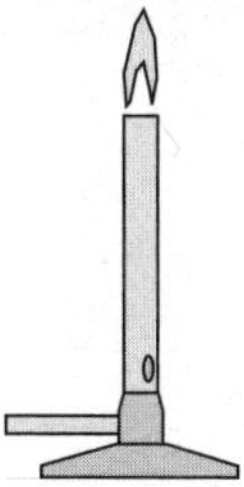	Produces a flame for heating.
Burette or Buret		Used for volumetric delivery of solutions. Used in titration experiment. Stopcock is small handle that controls the delivery of liquid.

Continued →

Name of Equipment	Picture	Function
Clamp and Ring Stand	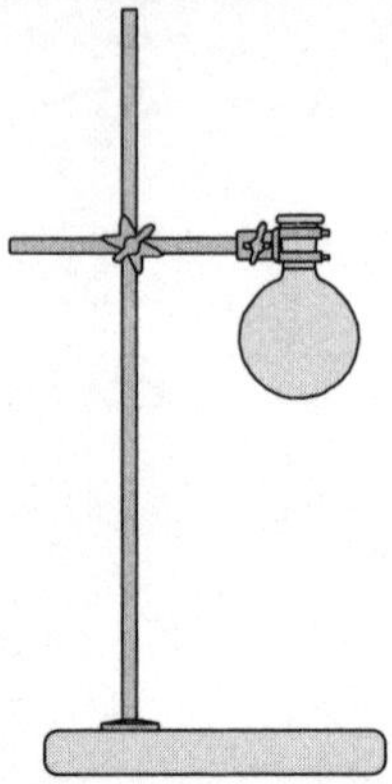	Holds funnels, flasks, wire gauze for burning.
Crucible		Heat resistant container used to heat compounds. Can be used in combination with ring stand and clamp.
Distillation Apparatus	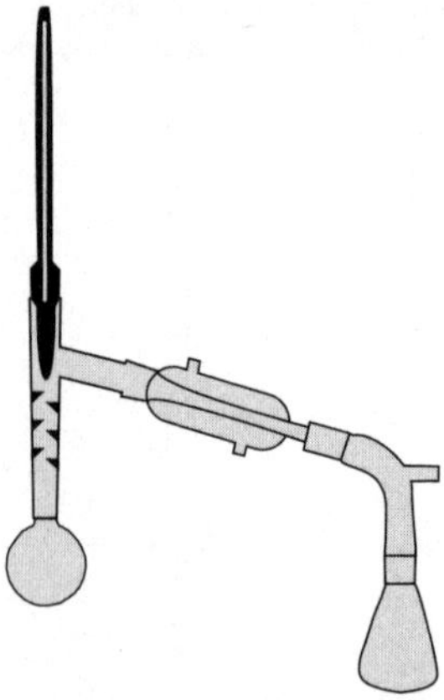	Used to separate a mixture of compounds based on boiling point.
Erlenmeyer Flask		Conical piece of lab equipment that is used for holding liquids. Not used for measuring.

Name of Equipment	Picture	Function
Evaporating Dish	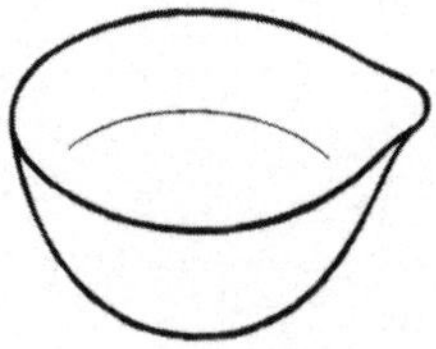	Used to hold aqueous solutions as they are heated in order to evaporate the water and leave the solid.
Filtering Flask	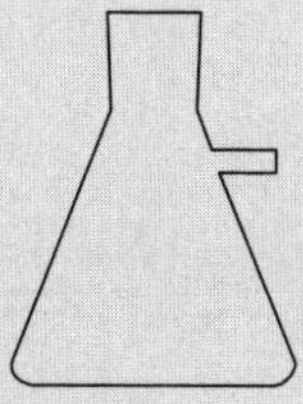	Used in combination with vacuum suction and Büchner funnel.
Funnel	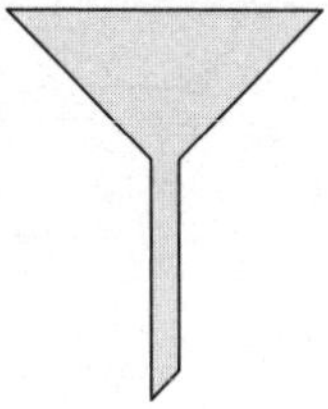	Along with filter paper can be used to separate solids from liquids.
Graduated Cylinder	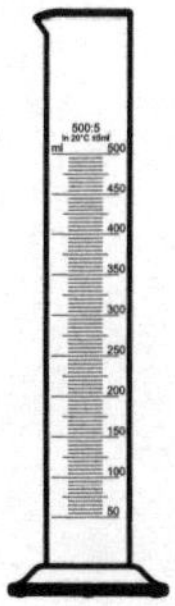	Used for measuring volumes of liquids. More accurate than a beaker and less accurate than volumetric equipment, such as pipets and burets.

Continued ⟶

Name of Equipment	Picture	Function
Hot Plate	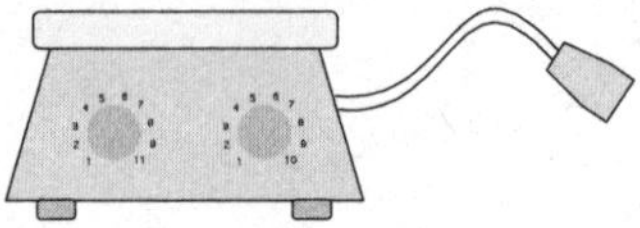	Electric device that allows for controlled delivery of heat. Often contains a metallic stirrer to allow for mixing.
Mortar and Pestle		Porcelain piece of equipment that can be used for crushing and grinding.
Separatory Funnel		Allows for the separation of immiscible liquids.
Thermometer	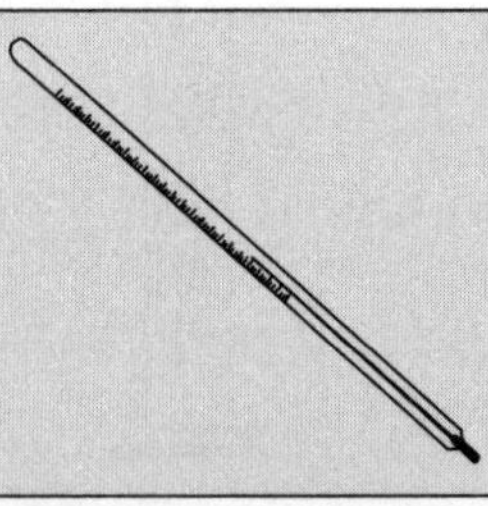	Used to measure temperature.
Well Plate	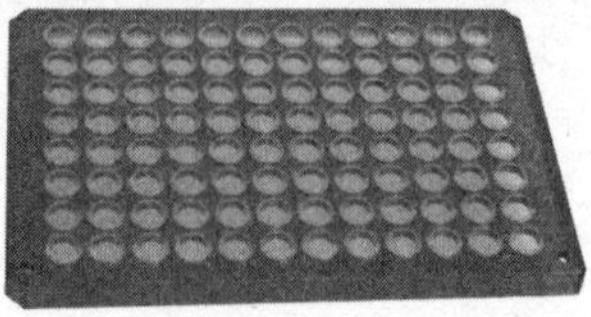	Provides small reaction wells that allow reactions to be carried out on a microscale.

Name of Equipment	Picture	Function
Volumetric Flask	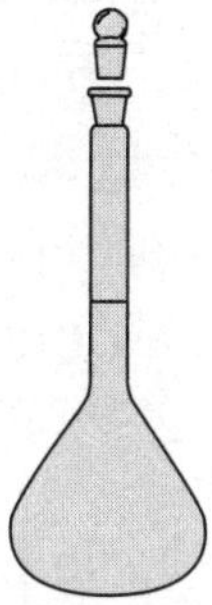	Used to accurately prepare solutions of various concentrations. Often used in titrations.
Volumetric Pipet	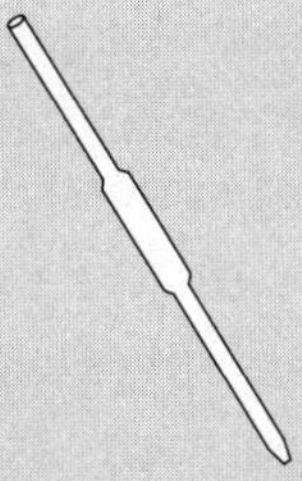	Used for the accurate measurement and transfer of liquid volumes. Often used during titrations and when diluting solutions.

B. Measurements, Accuracy, and Precision

1. Significant figures

   i. Determining significant figures. Any nonzero integers are always counted as significant.
      - ➤ 123 (3 sig fig.)
      - ➤ 5678.45 (6 sig fig.)
      - ➤ 222 (3 sig fig.)

   ii. Leading zeros are those that precede all of the nonzero digits and are never counted as significant.
      - ➤ 0.00022 (2 sig fig.)
      - ➤ 0.0235 (3 sig fig.)
      - ➤ 0.954 (3 sig fig.)

iii. Captive zeros are those that fall between nonzero digits and are counted as significant.
   - 10056 (5 sig fig.)
   - 2301 (4 sig fig.)
   - 100004 (6 sig fig.)

iv. Trailing zeros are those at the end of a number and are only significant if the number is written with a decimal point.
   - 100 (1 sig fig.)
   - 100. (3 sig fig.)
   - 100.000 (6 sig fig.)

v. In scientific notation, the 10^x part of the number is not counted as significant.
   - 1.2×10^5 (2 sig fig.)
   - 2.33×10^3 (3 sig fig.)
   - 1.2004×10^{-3} (5 sig fig.)

vi. Exact numbers have an unlimited number of significant figures. Exact numbers are those determined as a result of counting or by definition, e.g., 3 apples or 1.00 kg = 1000 g.

2. Significant figures in math operations

   i. When multiplying or dividing, limit the answer to the same number of significant figures that appear in the original data with the fewest number of significant figures.
      - $2.34 \div 1.34 = 1.75$ (rounded to 3 sig fig.)
      - $123 \times 6 = 700$ (rounded to 1 sig fig.)
      - $23 \div 34 = 0.68$ (rounded to 2 sig fig.)

   ii. When adding or subtracting, limit the answer to the same number of decimal places that appear in the original data with the fewest number of decimal places.
      - $57.3 + 12.6 = 69.9$ (1 decimal place)
      - $7.0 + 3.22 = 10.2$ (1 decimal place)
      - $11.3 - 6.66 = 4.6$ (1 decimal place)

3. Accuracy and precision
    i. Accuracy is how close a measurement is to the accepted (actual) value.
    ii. Precision is how close a series of measurements are to one another.
4. Weighing
    i. Allow cooling before weighing recently heated items, because convection currents may disturb the balance and hot items may damage a delicate electronic balance.
    ii. Use weighing boats to avoid damaging the pan of the balance with potentially corrosive chemicals.
5. Titrations
    i. Rinse pipets and burets with the solution that they will contain during the titration, not water (to avoid diluting the solution).
    ii. Read the buret scale at the point where the bottom of the meniscus is observed.
    iii. Record one uncertain figure, but no more. In burets that are graduated to a tenth of a milliliter, the final digit (second decimal place) should be recorded as either a "0" or a "5." Use "0" if the meniscus is directly ON a graduation, or "5" if it falls between two graduations. For example, if the bottom of the meniscus falls between 23.40 and 23.50, record 23.45.

V. Procedures

A. Methods of Separation

1. Filtration is the process of using a filter paper to separate an insoluble solid from a solution or liquid, such as filtering a precipitate from a solution.
2. Distillation is the process of heating a mixture of liquids and relying upon the difference in boiling point to collect one (the lower boiling-point component) above the other,

such as distillation of ethanol (boiling point 78°C) and water (boiling point 100°C).

3. Paper chromatography is the process of separating components based upon their relative affinities for a stationary phase (paper) and a moving phase (a solvent).

   i. R_f values are calculated by applying the following formula:

$$R_f = \frac{\text{distance traveled by component of mixture}}{\text{distance traveled by solvent front}}$$

VI. Safety

A. Safety Rules

1. Wear protective gear (goggles, aprons, gloves, closed-toe shoes, etc.).
2. Use a fume hood when appropriate (when working with volatile chemicals).
3. Dilute acids by *always adding acid to water*, not by adding water to acid.
4. Take care not to cross-contaminate chemicals.
5. Volatile and flammable substances should be heated using a water or oil bath and heating mantle/hot plate, not a naked flame.
6. Heat materials gently and point test tubes away from people.
7. Know the location of, and how to use, safety equipment.

VII. Data Analysis, Graphs, and Interpretation

A. Errors

1. Minimize by repeating experiments and measurements and averaging the results.

2. Percentage error can be calculated by applying the following formula:

$$\% \text{ Error} = \frac{|\text{Experimental value} - \text{Actual value}|}{\text{Actual value}} \times 100$$

B. Graphing

1. Place the independent variable (the variable you have control over and are changing) on the *x*-axis and the dependent variable (the variable you are measuring and is changing as a result of changing the independent variable) on the *y* axis.
2. Always label axes and place a title on the graph.

Test Tip

The contents of this section should be viewed as general laboratory knowledge and good practice when handling chemicals and data from experiments. They should be things that you have picked up from the lab component of your course. If you have taken a course with no lab component, or one where labs are not emphasized, you should study this chapter more closely.

Chapter 28

Writing Good Free-Response Answers

I. Use the Periodic Table Provided

Both the multiple-choice section and the free-response section of the AP Chemistry exam come with a periodic table that displays each element's atomic number, symbol, and average atomic mass. The names of the elements are not provided, so you should become familiar with the names of the common ones.

As you hopefully already know, the periodic table is a crucial tool to be used in any chemistry course and exam, and the AP exam is no different. You should find yourself referencing it many times during the exam, both for average atomic mass data and for information that will help you answer questions about patterns and positions such as periodicity.

II. Use the Equations and Constants Reference Sheet Provided

An Equations and Constants reference sheet is also provided for both the multiple-choice and the free-response sections of the AP Chemistry exam. There are two pages of reference material included.

The reference sheet is broken up into topics (Atomic Structure, Equilibrium, Kinetics, Gases, Liquids & Solutions, and Thermochemistry/Electrochemistry) that should make it easier to find the information that you may need. Check the Equations and Constants reference sheet to find the formula and constants that you already know, and use them to cross-reference the chapters in this book. You probably already know many of the formula and constants, so the reference sheet should be less intimidating than it might appear at first glance.

III. Present Your Work Neatly

AP readers are inclined to give the benefit of the doubt when grading the free-response questions to students that have presented their work in an orderly and neat fashion. You do not want the reader to have to look too hard for your answer, or for him or her to be wading through a messy answer. To help the reader (and therefore to help you), do the following;

1. The free-response questions will have multiple parts to them. Be sure your answers are labeled appropriately and accurately.
2. A picture can say a thousand words, and a quick sketch or diagram is a great way for you to show the reader that you know your stuff. Every sketch should be labeled with a few words or followed by a short explanation so the reader can make sense of what you have drawn.
3. In calculations, pay careful attention to significant figures and units. Although not necessary, circling or "boxing" your final numerical answers may help the reader. It is *not* absolutely necessary to use the factor label/dimensional analysis method in calculations either, but the method can be very helpful in ensuring that answers are presented neatly and that units are canceled out correctly.
4. Unlike some other AP exams, you do not have to write a formal essay for the AP Chemistry exam free-response questions. Limit your explanations to one or two sentences, and consider using bullet points or small lists to help make your answers clear and concise. This will be especially helpful if you are trying to outline an experimental procedure that will probably need to be carried out in specific sequence.
5. In the free-response questions, you may see some common key words that have a specific meaning. They include:

i. *Calculate* means that you need to provide a numerical answer to the question; show your work clearly, consider using dimensional analysis (the factor-label method), and always think about significant figures and units.

ii. *Draw* means that a diagram or sketch is necessary; consider adding a few words or labels to help with interpretation. Keep it simple. You do not have to be a professional artist—simple line drawings for laboratory equipment work well.

iii. *Explain/justify* requires you to provide a chemical concept or idea to clarify your answer.

iv. *Identify* requires you to pick out "a specific something" relevant to the question. Explanations are often required once an identification has been made.

v. *List* requires you to identify a series of "specific somethings" relevant to the question.

IV. Labs

Read and study Chapter 27 in this book to become familiar with the science practices, laboratory techniques, and the experiments that are relevant to the AP Chemistry course. A general knowledge of them is required, and you will certainly see questions relating to them on the exam.

It is also likely that you will be asked to design an experiment or experimental procedure to answer a specific scientific problem or question somewhere on the AP exam.

V. Graphical and Data Analysis

Although it is relatively unlikely that you would be required to actually graph data for one of the free-response questions, it is highly likely that somewhere on the exam you will be asked to interpret a graph or to interpret a collection of data. Since up to half of the multiple-choice questions could be presented in sets that have significant amounts of text or data associated with them, the multiple-choice section of the exam is a place where you will probably need to interpret data.

Here are some simple pointers that will help you deal with any question relating to graphs and data.

1. Graph setup and plotting

   Generally, independent variables appear on the *x*-axis and dependent variables appear on the *y*-axis.

2. Using the graph to answer further questions

   Once the data are plotted on the graph, you may need to use the data to answer a series of questions from the plot. These answers can be quick mathematical calculations involving slope or simple extrapolations of the data.

3. Look for patterns and/or exceptions in sets of data. Relate data to physical and chemical properties. For example, melting and boiling points being related to intermolecular forces and intra bonding, or PES data relating to electronic configurations.

PERIODIC TABLE OF THE ELEMENTS

1 **H** 1.008																	2 **He** 4.00
3 **Li** 6.94	4 **Be** 9.01											5 **B** 10.81	6 **C** 12.01	7 **N** 14.01	8 **O** 16.00	9 **F** 19.00	10 **Ne** 20.18
11 **Na** 22.99	12 **Mg** 24.30											13 **Al** 26.98	14 **Si** 28.09	15 **P** 30.97	16 **S** 32.06	17 **Cl** 35.45	18 **Ar** 39.95
19 **K** 39.10	20 **Ca** 40.08	21 **Sc** 44.96	22 **Ti** 47.90	23 **V** 50.94	24 **Cr** 52.00	25 **Mn** 54.94	26 **Fe** 55.85	27 **Co** 58.93	28 **Ni** 58.69	29 **Cu** 63.55	30 **Zn** 65.39	31 **Ga** 69.72	32 **Ge** 72.59	33 **As** 74.92	34 **Se** 78.96	35 **Br** 79.90	36 **Kr** 83.80
37 **Rb** 85.47	38 **Sr** 87.62	39 **Y** 88.91	40 **Zr** 91.22	41 **Nb** 92.91	42 **Mo** 95.94	43 **Tc** (98)	44 **Ru** 101.1	45 **Rh** 102.91	46 **Pd** 106.42	47 **Ag** 107.87	48 **Cd** 112.41	49 **In** 114.82	50 **Sn** 118.71	51 **Sb** 121.75	52 **Te** 127.60	53 **I** 126.91	54 **Xe** 131.29
55 **Cs** 132.91	56 **Ba** 137.33	57 ***La** 138.91	72 **Hf** 178.49	73 **Ta** 180.95	74 **W** 183.85	75 **Re** 186.21	76 **Os** 190.2	77 **Ir** 192.2	78 **Pt** 195.08	79 **Au** 196.97	80 **Hg** 200.59	81 **Tl** 204.38	82 **Pb** 207.2	83 **Bi** 208.98	84 **Po** (209)	85 **At** (210)	86 **Rn** (222)
87 **Fr** (223)	88 **Ra** 226.02	89 **†Ac** 227.03	104 **Rf** (261)	105 **Db** (262)	106 **Sg** (266)	107 **Bh** (264)	108 **Hs** (277)	109 **Mt** (268)	110 **Ds** (271)	111 **Rg** (272)							

*Lanthanide Series	58 **Ce** 140.12	59 **Pr** 140.91	60 **Nd** 144.24	61 **Pm** (145)	62 **Sm** 150.4	63 **Eu** 151.97	64 **Gd** 157.25	65 **Tb** 158.93	66 **Dy** 162.50	67 **Ho** 164.93	68 **Er** 167.26	69 **Tm** 168.93	70 **Yb** 173.04	71 **Lu** 174.97
†Actinide Series	90 **Th** 232.04	91 **Pa** 231.04	92 **U** 238.03	93 **Np** (237)	94 **Pu** (244)	95 **Am** (243)	96 **Cm** (247)	97 **Bk** (247)	98 **Cf** (251)	99 **Es** (252)	100 **Fm** (257)	101 **Md** (258)	102 **No** (259)	103 **Lr** (262)

AP Chemistry Equations and Constants

Throughout the test the following symbols have the definitions specified unless otherwise noted.

L, mL	= liter(s), milliliter(s)	mm Hg	= millimeters of mercury
g	= gram(s)	J, kJ	= joule(s), kilojoule(s)
nm	= nanometer(s)	V	= volt(s)
atm	= atmosphere(s)	mol	= mole(s)

ATOMIC STRUCTURE

$E = h\nu$

$c = \lambda\nu$

E = energy

ν = frequency

λ = wavelength

Planck's constant, $h = 6.626 \times 10^{-34}$ J s

Speed of light, $c = 2.998 \times 10^{8}$ m s^{-1}

Avogadro's number $= 6.022 \times 10^{23}$ mol^{-1}

Electron charge, $e = -1.602 \times 10^{-19}$ Coulomb

EQUILIBRIUM

$$K_c = \frac{[C]^c[D]^d}{[A]^a[B]^b}, \text{ where } a\,A + b\,B \rightleftarrows c\,C + d\,D$$

$$K_p = \frac{(P_C)^c(P_D)^d}{(P_A)^a(P_B)^b}$$

$$K_a = \frac{[H^+][A^-]}{[HA]}$$

$$K_b = \frac{[OH^-][HB^+]}{[B]}$$

$$K_w = [H^+][OH^-] = 1.0 \times 10^{-14} \text{ at } 25°C$$

$$= K_a \times K_b$$

$$pH = -\log[H^+],\ pOH = -\log[OH^-]$$

$$14 = pH + pOH$$

$$pH = pK_a + \log\frac{[A^-]}{[HA]}$$

$$pK_a = -\log K_a,\ pK_b = -\log K_b$$

Equilibrium Constants

K_c (molar concentrations)

K_p (gas pressures)

K_a (weak acid)

K_b (weak base)

K_w (water)

KINETICS

$$\ln[A]_t - \ln[A]_0 = -kt$$

$$\frac{1}{[A]_t} - \frac{1}{[A]_0} = kt$$

$$t_{1/2} = \frac{0.693}{k}$$

k = rate constant

t = time

$t_{1/2}$ = half-life

GASES, LIQUIDS, AND SOLUTIONS

$$PV = nRT$$

$$P_A = P_{\text{total}} \times X_A, \text{ where } X_A = \frac{\text{moles A}}{\text{total moles}}$$

$$P_{total} = P_A + P_B + P_C + \ldots$$

$$n = \frac{m}{\boldsymbol{M}}$$

$$\text{K} = {}^\circ\text{C} + 273$$

$$D = \frac{m}{V}$$

$$KE \text{ per molecule} = \frac{1}{2}mv^2$$

Molarity, M = moles of solute per liter of solution

$$A = abc$$

P = pressure
V = volume
T = temperature
n = number of moles
m = mass
$\boldsymbol{M}$ = molar mass
D = density
KE = kinetic energy
v = velocity
A = absorbance
a = molar absorptivity
b = path length
c = concentration

Gas constant, $R = 8.314 \text{ J mol}^{-1}\text{ K}^{-1}$
$= 0.08206 \text{ L atm mol}^{-1}\text{ K}^{-1}$
$= 62.36 \text{ L torr mol}^{-1}\text{ K}^{-1}$
1 atm = 760 mm Hg
= 760 torr
STP = 0.00°C and 1.000 atm

THERMOCHEMISTRY/ ELECTROCHEMISTRY

$$q = mc\Delta T$$

$$\Delta S^\circ = \sum S^\circ \text{ products} - \sum S^\circ \text{ reactants}$$

$$\Delta H^\circ = \sum \Delta H_f^\circ \text{ products} - \sum \Delta H_f^\circ \text{ reactants}$$

$$\Delta G^\circ = \sum \Delta G_f^\circ \text{ products} - \sum \Delta G_f^\circ \text{ reactants}$$

$$\Delta G^\circ = \Delta H^\circ - T\Delta S^\circ$$
$$= -RT \ln K$$
$$= -nFE^\circ$$

$$I = \frac{q}{t}$$

q = heat
m = mass
c = specific heat capacity
T = temperature
S° = standard entropy
H° = standard enthalpy
G° = standard free energy
n = number of moles
E° = standard reduction potential
I = current (amperes)
q = charge (coulombs)
t = time (seconds)

Faraday's constant, F = 96,485 Coulombs per mole of electrons

$$1 \text{ volt} = \frac{1 \text{ Joule}}{1 \text{ Coulomb}}$$

Notes

Notes

Notes

Notes

Notes

Notes